Weapons of Mass Destruction

Detection of Nuclear Weapons and Materials

WEAPONS OF MASS DESTRUCTION

Proliferation Control Regimes
Adelfo Gaspari (Editor)
2009. ISBN: 978-1-60741-115-4

Nuclear Weapons' Role in 21st Century U.S Policy
Dominick R. Pelligrini (Editor)
2010. ISBN: 978-1-60741-478-0

China and Proliferation of Weapons of Mass Destruction and Missiles: Policy Issues
Shirley A. Kan
2010. ISBN: 978-1-60741-248-9

Detection of Nuclear Weapons and Materials
Pietro Egidi (Editor)
2010. ISBN: 978-1-60741-512-1

North Korea's Nuclear and Ballistic Weapons
Simon R. Holden (Editor)
2010. ISBN: 978-1-60876-844-8

WEAPONS OF MASS DESTRUCTION

DETECTION OF NUCLEAR WEAPONS AND MATERIALS

PIETRO EGIDI
EDITOR

Nova Science Publishers, Inc.

New York

For permission to use material from this book please contact us:
Telephone 631-231-7269; Fax 631-231-8175
Web Site: http://www.novapublishers.com

LIBRARY OF CONGRESS CATALOGING-IN-PUBLICATION DATA

Detection of nuclear weapons and materials / editor, Pietro Egidi.
p. cm.
Includes index.
ISBN 978-1-60741-512-1 (hardcover)
1. Radioactive substances--Detection. 2. Nuclear weapons. 3. Nuclear terrorism--United States--Prevention. 4. Radiation warning systems. 5. Nuclear nonproliferation--Government policy--United States. 6. Nuclear arms control--Verification. 7. Smuggling--Prevention. I. Egidi, Pietro.
TK9185.D48 2009
363.325'563--dc22

2009041942

Published by Nova Science Publishers, Inc. ✢ New York

CONTENTS

PREFACE

Detection of nuclear weapons and special nuclear material (SNM) is crucial to thwarting nuclear proliferation and terrorism and to securing weapons and materials worldwide. Congress has funded a portfolio of detection R&D and acquisition programs, and has mandated inspection at foreign ports of all U.S.-bound cargo containers using two types of detection equipment. Nuclear weapons contain SNM, which produces unique or suspect signatures that can be detected. It emits radiation, notably gamma rays (high-energy photons) and neutrons. SNM is very dense, so it produces a bright image on a radiograph when X-rays or gamma rays are beamed through a container in which it is hidden. Some detection technology is advancing faster than many have expected. It is easier and less costly to accelerate a program in R&D than in production. This book emphasizes the ongoing improvement in detection capabilities which produce uncertainties for terrorists that will increase over time, adding deterrence beyond that of the capabilities themselves.

This book consists of public documents which have been located, gathered, combined, reformatted, and enhanced with a subject index, selectively edited and bound to provide easy access.

Chapter 1 - The Department of Homeland Security's (DHS) Domestic Nuclear Detection Office (DNDO) is responsible for addressing the threat of nuclear smuggling. Radiation detection portal monitors are part of the U.S. defense against such threats. In 2007, Congress required that funds for new advanced spectroscopic portal (ASP) monitors could not be spent until the Secretary of DHS certified that these machines represented a significant increase in operational effectiveness over currently deployed portal monitors. In addition to other tests, DNDO conducted the Phase 3 tests on ASPs to identify areas in which the ASPs needed improvement. GAO was asked to assess (1) the degree to which the Phase 3 test report accurately depicts the test results and (2) the appropriateness of using the Phase 3 test results to determine whether ASPs represent a significant improvement over current radiation detection equipment. GAO also agreed to provide its observations on special tests conducted by Sandia National Laboratories (SNL).

Chapter 2 - Detection of nuclear weapons and special nuclear material (SNM, certain types of uranium and plutonium) is crucial to thwarting nuclear proliferation and terrorism and to securing weapons and materials worldwide. Congress has funded a portfolio of detection R&D and acquisition programs, and has mandated inspection at foreign ports of all U.S.-bound cargo containers using two types of detection equipment.

Nuclear weapons contain SNM, which produces unique or suspect signatures that can be detected. It emits radiation, notably gamma rays (high-energy photons) and neutrons. SNM is very dense, so it produces a bright image on a radiograph (a picture like a medical x-ray) when x-rays or gamma rays are beamed through a container in which it is hidden. Using lead or other shielding to attenuate gamma rays would make this image larger. Nuclear weapons produce detectable signatures, such as radiation generated by or a noticeable image on a radiograph. Other detection techniques are also available.

Nine technologies illustrate the detection portfolio: (1) A new scintillator material to improve detector performance and lower cost. (2) GADRAS, an algorithm to determine the materials in a container by analyzing gamma-ray spectra. If materials are the "eyes and ears" of detectors, algorithms are the "brains." (3) A project to simulate large numbers of experiments in order to improve detection system performance. (4, 5) Two Cargo Advanced Automated Radiography Systems (CAARS) to detect high-density material based on the principle that it becomes less transparent to photons of higher energy, unlike other material. (6) A third CAARS to detect material with high atomic number (Z, number of protons in an atom's nucleus) based on the principle that Z affects how material scatters photons. (7) A system to generate a 3-D image of the contents of a container based on the principle that Z and density strongly affect the degree to which muons (a type of subatomic particle) scatter. (8) Nuclear resonance fluorescence imaging to identify materials based on the spectrum of gamma rays a nucleus emits when struck by photons of a specific energy. (9) The Photonuclear Inspection and Threat Assessment System to detect SNM up to 1 km away, unlike other systems that operate at very close range. It would beam high-energy photons at distant targets to stimulate fission in SNM, producing characteristic signatures that may be detected.

This analysis leads to several observations for Congress. Some detection technology is advancing faster than many have expected. It is easier and less costly to accelerate a program in R&D than in production. "Concept of operations" is crucial to detection system effectiveness. Congress may wish to address gaps and synergisms in the technology portfolio. Congress need not depend solely on procedures developed by executive agencies to test detection technologies, but may specify tests an agency is to conduct. Ongoing improvement in detection capabilities produces uncertainties for terrorists that will increase over time, adding deterrence beyond that of the capabilities themselves. This report will be updated occasionally.

Congress has been interested in detecting nuclear weapons and the special materials needed to make them for many years, especially since 9/11. Nuclear detection has many applications for countering nuclear terrorism and nuclear proliferation, such as securing nuclear weapons and materials in U.S., Russian, and other nuclear facilities, tracking materials at border crossings and choke points, screening maritime cargo containers, and examining actual or suspect nuclear sites.

The United States currently uses several types of nuclear detection equipment. All have significant shortcomings. Some work only at very short range; some cannot identify the material emitting radiation, which can lead to false alarms and interrupt commerce; some depend on operator skill, and may be defeated by a clever smuggler or a sleepy operator; and some are easily defeated by shielding. In an effort to overcome such problems, Congress has funded a pipeline of advanced-technology research, development, and acquisition.

This report seeks to help Congress understand this technology. It discusses the science of detecting nuclear weapons and materials, describes nine advanced U.S. technologies selected to illustrate the range of projects in the pipeline, and offers observations for Congress. The report does not compare technologies.[1] The inclusion of the nine technologies should not be taken to mean that CRS judges them to be better than the hundreds of others not considered here. The report does not discuss the controversial Advanced Spectroscopic Portal because a detailed discussion of it could draw attention from the other technologies considered here.

The scope of this report excludes the organization of the government for dealing with nuclear detection, the role of intelligence and law enforcement in detecting terrorist nuclear weapons, detection of radiological dispersal devices (such as "dirty bombs"), the role of nuclear forensics in attributing an attack to its perpetrator, response to a nuclear attack, and the architecture of a national nuclear detection system.[2] Nor does it discuss possible means by which terrorists might acquire a bomb, or whether they could make a bomb on their own. Much has been written on these topics.[3] While many who are concerned with nuclear detection focus on thwarting nuclear terrorism, this report focuses on technology per se. It avoids extensive discussion of means to counter detection to avoid classified information.

Nuclear detection technology has a dual role in thwarting a terrorist nuclear attack—deterrence and defense. Deterrence means dissuasion from an action by threat of unacceptable consequences. Terrorists may be deterred from a nuclear strike by one of the few consequences unacceptable to them: failure. Detection systems would raise that risk. These systems could also make a terrorist nuclear strike too complex to succeed. But other factors would also have these effects: the difficulty of fabricating a bomb, the chance that law enforcement or intelligence would detect efforts to obtain a bomb, the possible inability to detonate a purloined bomb, and the risk that scientists recruited for the plot would defect. Such risks would disappear, however, if terrorists were given a bomb and operating instructions. They would then only need to mount a smuggling operation. In that case, the role of nuclear detection systems would change: they would become the main defense.

Chapter 3 - The U.S. government has implemented a series of programs to protect the nation against terrorist nuclear attack. Some of these programs predate September 11, 2001, while others were established since then. Most programs are within the Nuclear Regulatory Commission; the Departments of Defense, Energy, and State; and agencies that became part of the Department of Homeland Security (DHS) upon its creation, and they are focused on detecting the illicit acquisition and shipment of nuclear and radiological materials and protecting and securing nuclear weapons. These disparate programs have historically been viewed as lacking coordination and centralized oversight.

In 2005, the Domestic Nuclear Detection Office (DNDO) was established within the Department of Homeland Security to centralize coordination of the federal response to an unconventional nuclear threat. The office was codified in 2006 through the passage of the SAFEPort Act (P.L. 109-347) and given specific statutory responsibilities to protect the United States against radiological and nuclear attack, including the responsibility to develop a "global nuclear detection architecture." Determining the range of existing federal efforts protecting against nuclear attack, coordinating the outcomes of these efforts, identifying overlaps and gaps between them, and integrating the results into a single architecture are likely to be evolving, ongoing tasks.

The global nuclear detection architecture is a multi-layered system of detection technologies, programs, and guidelines designed to enhance the nation's ability to detect and prevent a radiological or nuclear attack. Among its components are existing programs in nuclear detection operated by other federal agencies and new programs put into place by DNDO. The global nuclear detection architecture is developed by DNDO in coordination with other federal agencies implementing nuclear detection efforts and this coordination is essential to the success of the architecture.

This architecture is a complicated system of systems. Measuring the success of the architecture relative to its individual components and the effectiveness of additional investments are challenges. The DNDO is developing risk and cost methodologies to be applied to the architecture in order to understand and prioritize the various nuclear detection programs and activities in multiple federal agencies.

Congress, in its oversight capacity, has shown interest in the development and implementation of the global nuclear detection architecture and in the decision-making process attendant to investments in it. Other issues that may be foci of congressional attention include the balance between investment in near-term and long-term solutions for architecture gaps, the degree and efficacy of federal agency coordination, the mechanism for setting agency investment priorities in the architecture, and the efforts DNDO has undertaken to retain institutional knowledge regarding this sustained architecture effort.

Chapter 4 - In April 2005, a Presidential Directive established the Domestic Nuclear Detection Office (DNDO) within the Department of Homeland Security to enhance and coordinate federal, state, and local efforts to combat nuclear smuggling domestically and abroad. DNDO was directed to develop, in coordination with the departments of Defense (DOD), Energy (DOE), and State (State), an enhanced global nuclear detection architecture—an integrated system of radiation detection equipment and interdiction activities. DNDO implements the domestic portion of the architecture, while DOD, DOE, and State are responsible for related programs outside the U.S.

This testimony provides preliminary observations based on ongoing work addressing (1) the status of DNDO's efforts to develop a global nuclear detection architecture, (2) the challenges DNDO and other federal agencies face in implementing the architecture, and (3) the costs of the programs that constitute the architecture. This statement draws on prior GAO reviews of programs constituting the architecture, and GAO's work on strategic planning.

In: Detection of Nuclear Weapons and Materials
Editor: Pietro Egidi
ISBN: 978-1-60741-512-1

Chapter 1

COMBATING NUCLEAR SMUGGLING: DHS'S PHASE 3 TEST REPORT ON ADVANCED PORTAL MONITORS DOES NOT FULLY DISCLOSE THE LIMITATIONS OF THE TEST RESULTS

Government Accountability Office

WHY GAO DID THIS STUDY

The Department of Homeland Security's (DHS) Domestic Nuclear Detection Office (DNDO) is responsible for addressing the threat of nuclear smuggling. Radiation detection portal monitors are part of the U.S. defense against such threats. In 2007, Congress required that funds for new advanced spectroscopic portal (ASP) monitors could not be spent until the Secretary of DHS certified that these machines represented a significant increase in operational effectiveness over currently deployed portal monitors. In addition to other tests, DNDO conducted the Phase 3 tests on ASPs to identify areas in which the ASPs needed improvement. GAO was asked to assess (1) the degree to which the Phase 3 test report accurately depicts the test results and (2) the appropriateness of using the Phase 3 test results to determine whether ASPs represent a significant improvement over current radiation detection equipment. GAO also agreed to provide its observations on special tests conducted by Sandia National Laboratories (SNL).

WHAT GAO RECOMMENDS

GAO's recommendations include proposing that the Secretary of DHS revise the Phase 3 report to better disclose test results and limitations if it is to be used in any certification decision for ASP acquisition. DHS disagreed with two of GAO's recommendations but agreed to take action on a third. GAO continues to believe that all of its recommendations need to be implemented.

WHAT GAO FOUND

Because the limitations of the Phase 3 test results are not appropriately stated in the Phase 3 test report, the report does not accurately depict the results from the tests and could potentially be misleading. In the Phase 3 tests, DNDO performed a limited number of test runs. Because of this, the test results provide little information about the actual performance capabilities of the ASPs. The report often presents each test result as a single value; but considering the limited number of test runs, the results would be more appropriately stated as a range of potential values. For example, the report narrative states in one instance that an ASP could identify a source material during a test 50 percent of the time. However, the narrative does not disclose that, given the limited number of test runs, DNDO can only estimate that the ASP would correctly identify the source from about 15 percent to about 85 percent of the time—a result that lacks the precision implied by DNDO's narrative. DNDO's reporting of the test results in this manner makes them appear more conclusive and precise than they really are. The purpose of the Phase 3 tests was to conduct a limited number of test runs in order to identify areas in which the ASP software needed improvement. While aspects of the Phase 3 report address this purpose, the preponderance of the report goes beyond the test's original purpose and makes comparisons of the performance of the ASPs with one another or with currently deployed portal monitors.

In GAO's view, it is not appropriate to use the Phase 3 test report in determining whether the ASPs represent a significant improvement over currently deployed radiation equipment because the limited number of test runs do not support many of the comparisons of ASP performance made in the Phase 3 report. As the report shows, if an ASP can identify a source material every time during a test, but the test is run only five times, the only thing that can be inferred with a high level of statistical confidence is that the probability of identification is no less than about 60 percent. Although DNDO states in the Phase 3 test report that the results will be relevant to the Secretary's certification that the ASPs represent a significant increase in operational effectiveness, it does not clarify in what ways the results will be relevant. Furthermore, DNDO offers no explanation as to why it changed its view from the Phase 3 test plan, which states that these tests will not be used to support a certification decision.

The goal of SNL's special tests was, among other things, to identify potential vulnerabilities in the ASPs by using different test scenarios from those that DNDO planned to use in other ASP tests. SNL concluded in its test report that the ASPs' software and hardware can be improved and that rigor could be added to DNDO's testing methods. Furthermore, the report acknowledges that (1) a specific objective of the testing at the Nevada Test Site was to refine and improve the ASP's performance and (2) the special tests were never intended to demonstrate conformity of the ASPs with specific performance requirements. In GAO's view, these statements appear to accurately describe the purpose, limitations, and results of the special tests.

ABBREVIATIONS

ASP	advanced spectroscopic portal
CBP	Customs and Border Protection

DHS	Department of Homeland Security
DNDO	Domestic Nuclear Detection Office
PVT	polyvinyl toluene
RIID	radioactive isotope identification device
SNL	Sandia National Laboratories

September 30, 2008

The Honorable John D. Dingell
Chairman
The Honorable Joe Barton
Ranking Member
Committee on Energy and Commerce
House of Representatives

The Honorable Bart Stupak
Chairman
The Honorable John Shimkus
Ranking Member
Subcommittee on Oversight and Investigations
Committee on Energy and Commerce
House of Representatives

Preventing a nuclear weapon or radiological dispersal device (a "dirty bomb") from being smuggled into the United States is a key national security priority. The Department of Homeland Security (DHS), through its Domestic Nuclear Detection Office (DNDO), has lead responsibility for conducting the research, development, testing, and evaluation of equipment that can be used to detect smuggled nuclear or radiological materials. U.S. Customs and Border Protection (CBP) is responsible for screening cargo as it enters the nation at our borders, including operating radiation detection equipment to intercept dangerous nuclear and radiological materials.

Much of DNDO's work on radiation detection equipment has focused on the development and use of radiation detection portal monitors that can screen vehicles, people, and cargo entering the United States. In the case of cargo, these portal monitors typically include two detector panels that form sidewalls of a portal through which trailer trucks carrying cargo containers must pass. Currently, CBP employs portal monitors made of polyvinyl toluene (a plastic), known as PVTs, which can detect the presence of radiation but cannot identify the specific radiological material generating the radiation. As a result, PVTs cannot distinguish between benign, naturally occurring radiological materials such as ceramic tile, and dangerous materials such as highly enriched uranium. Therefore, if a PVT detects the presence of radiation in a shipping container during a primary inspection, CBP conducts a second inspection with another PVT and uses a handheld radioactive isotope identification device (RIID) to identify the type of source material emitting radiation. However, RIIDs use detectors that are relatively small and, as a result, are limited in their ability to correctly identify radiological and nuclear source materials, so CBP officials sometimes must consult with scientists at CBP's Laboratories and Scientific Services or physically search a container to identify the radiation source. Nonetheless, CBP officials have stated that the current system of radiation detection equipment and standard operating procedures provide the best possible radiological and nuclear screening coverage available with current technology and that it does not have a significant adverse impact on the flow of commerce.

In 2005, in an effort to overcome the technical limitations of PVTs and RIIDs, DNDO sponsored research, development, and testing intended to produce Advanced Spectroscopic Portal (ASP) monitors, which are designed to both detect the radiological or nuclear source material and identify the specific type of source material. According to DNDO, ASPs will reduce the number of false positives or nuisance alarms—instances in which a portal monitor detects the presence of radiation and signals an alarm, but the source material turns out to be benign.

In July 2006, DNDO awarded contracts worth potentially \$1.2 billion to three vendors to further develop and produce ASPs over five years. Shortly thereafter, Congress required in DHS's fiscal year 2007 appropriation that the Secretary of Homeland Security certify that ASPs represent "a significant increase in operational effectiveness" before DNDO could order full-scale production. Congress enacted a similar requirement in DHS's fiscal year 2008 appropriation.

In early 2007, DNDO conducted formal tests on ASPs in two phases at the Nevada Test Site. The first, called Phase 1, was designed to assess the ASP's performance capabilities in order to support a full-scale production decision with a high degree of statistical confidence. DNDO told us on multiple occasions and in a written response that only data collected during Phase 1 would be included in the final report presented to the Secretary to request ASP certification. According to DNDO, the second group of tests, called Phase 3, provided data for algorithm (software) improvement that targeted specific and known areas in need of work and data to aid in the development of secondary screening operations and procedures.[1] For example, the Phase 3 tests attempt to determine, among other things, how ASP performance is affected by the presence of various substances—known as shielding materials—that partially absorb or alter the radiation emitted by the source material in a container. During the Phase 3 tests, DNDO also tested the ASPs to determine how their performance changes when the container moves through the portal monitor at different speeds.

The size of the samples used in the Phase 1 and Phase 3 tests are important in determining the confidence one can place in the test results.[2] Larger sample sizes, such as the 15 to 60 test runs performed during the Phase 1 tests, usually allow a more precise interpretation of results, i.e., estimates of ASP performance may fall within a fairly narrow range of values. Conversely, estimates drawn from small sample sizes, such as the 1 to 10 test runs done for Phase 3, normally have much less precision associated with them—thus the range of potential values may be quite wide. According to DNDO's Phase 3 test plan, the Phase 3 testing consisted of small sample sizes to allow more time to test a wide range of source materials in order to make improvements to the ASPs. However, the small samples associated with the Phase 3 tests would make it difficult to use the test results as a reliable indicator of the ASPs' performance capabilities or for comparisons of performance among various detection systems. In contrast, the Phase 1 tests involved larger sample sizes so that DNDO could assess the performance capabilities of the ASPs with a higher degree of statistical precision. Because of the small sample sizes, the Phase 3 test plan stated that Phase 3 testing would not be used to support a full-scale production decision.

In September 2007, we testified that, in our view, DNDO's Phase 1 tests did not represent an objective or rigorous assessment of the ASPs. More specifically, we stated that DNDO used biased test methods that enhanced the apparent performance of the ASPs and did not identify the limitations of the ASPs' detection capabilities.[3] During that hearing, DNDO's Director changed DNDO's position and stated that the Secretary could use the Phase 3 tests in combination with other test results when deciding whether to certify the ASPs.

In its report on the Phase 3 tests, DNDO states that

- The ASPs were as good as or better than the PVTs at detecting the presence of radiological source materials at low levels of radiological activity.
- The performance of the ASPs from each of the three vendors was statistically indistinguishable with few exceptions, for each category of source material (i.e., medical sources, industrial sources, and special nuclear material).
- The performance of the RIIDs was poor compared to the performance of the ASPs in identifying the specific radiological or nuclear source material inside a container.

At the same time the Phase 1 and Phase 3 tests were ongoing, Sandia National Laboratories (SNL), at the request of DNDO, conducted a series of tests that would go beyond those covered in the Phase 1 and Phase 3 tests. The goal of these tests, called special tests, was, among other things, to identify potential vulnerabilities in the ASPs by using source materials and test scenarios different from those that DNDO planned to use in either the Phase 1 or the Phase 3 tests. The tests were "blind" in that neither the ASP vendors nor the ASP test operators knew what was in the containers being tested. The review was also to focus on vulnerabilities in the test processes DNDO described in its Phase 1 and Phase 3 test plans.

Given DNDO's change in how it believes the Phase 3 test results may be applied, the significant costs of the ASPs, and the importance of protecting our borders from nuclear smuggling you asked us to assess (1) the degree to which the Phase 3 test report accurately depicts the test results and (2) the appropriateness of using the Phase 3 test results to determine whether ASPs represent a significant improvement over current radiation detection equipment. We also agreed to provide our observations on the special tests conducted by SNL.

To perform our work, we reviewed the Phase 3 test report and SNL's special tests report. We met with key officials from the National Institutes of Standards and Technology who were responsible for designing part of the Phase 3 tests and analyzing their results. We also relied on documents and other evidence gathered during our previous review of ASP testing at the Nevada Test Site. We conducted this performance audit from February 2008 to August 2008 in accordance with generally accepted government auditing standards. Those standards require that we plan and perform the audit to obtain sufficient, appropriate evidence to provide a reasonable basis for our findings and conclusions based on our audit objectives. We believe that the evidence obtained provides a reasonable basis for our findings and conclusions based on our audit objectives.

RESULTS IN BRIEF

Because the limitations of the Phase 3 test results are not appropriately stated in the Phase 3 test report, the report does not accurately depict the results from the tests and could be misleading. In the Phase 3 tests, DNDO performed a limited number of test runs. Because of this, the test results provide little information about the actual performance capabilities of the ASPs. The report narrative often presents test results using a single value or percentage. Considering the limited number of test runs, the results would be more appropriately stated as a range of potential values. This important limitation is apparent only by reviewing more technical data elsewhere in the report. For example, the report narrative states in one instance that an ASP could identify a source material during a specific test 50 percent of the time. However, the narrative does not disclose that, given the limited number of test runs, DNDO can only estimate that the ASP would be able to correctly identify the source from about 15 percent to about 85 percent of the time—a result that lacks the precision implied by DNDO's narrative. DNDO's reporting of the test results in this manner makes the results appear more conclusive and precise than they really are. For example, the executive summary states that the ASPs "demonstrated detection limits equal to or better than those of any of the PVT systems," but fails to mention that the test used only a single source material and that the results would not necessarily be the same for other radiological sources. In fact, the Phase 3 results showed that the PVTs could detect some source materials much better than the ASPs. The purpose of the Phase 3 tests was to identify areas in which the ASP software needed improvement. While aspects of the Phase 3 report address this purpose, the preponderance of the report goes beyond the test's original purpose and makes comparisons of the performance of the ASPs with each other or with PVTs and RIIDs.

In our view, it is not appropriate to use the Phase 3 test report in determining whether the ASPs represent a significant improvement over currently deployed radiation equipment because the limited number of test runs does not support many of the comparisons of ASP performance made in the Phase 3 report. As the report shows, if an ASP can identify a source material every time during a test, but the test is run only five times, the only thing that can be inferred at the 9 percent confidence level is that the true probability of identification is no less than approximately 60 percent. Although DNDO states in the Phase 3 test report that the test results will be relevant to the Secretary's certification that the ASPs represent a significant increase in operational effectiveness, the report does not clarify in what ways the results will be relevant. Furthermore, DNDO offers no explanation as to why it changed its view from the Phase 3 test plan which states that the Phase 3 tests will not be used to support a certification decision.

Regarding SNL's special tests, SNL concluded in its test report that the ASPs' software and hardware can be improved in some areas and that rigor could be added to DNDO's testing methods. Furthermore, the report acknowledges that (1) a specific objective of the testing at the Nevada Test Site was to refine and improve the ASP software performance and (2) the special tests "were never intended to demonstrate conformity of the [ASP] systems against specific performance requirements." In our view, these statements appear to accurately describe the purpose, limitations, and results of the special tests. In addition, the report concluded that upon reviewing data from DNDO's previous ASP tests, the reported results were consistent with the data that DNDO collected and that as a result, "DNDO's ASP system assessment was not biased." It is important to note, however, SNL's report also concluded that the "ASP system assessment [in 2007] was not biased" and that SNL "observed no data suggesting that the ASP system performance was inappropriately manipulated." In making this statement, SNL is referring to the data derived from ASP tests. In contrast, when we stated in September 2007 that DNDO's Phase 1 tests were biased, we were referring to the test methods DNDO used, such as (1) using the same test sources and shielding materials during preliminary runs as were used during the actual tests, and (2) not using standard CBP operating procedures in testing the RIIDs.

We are recommending that the Secretary of DHS use the results of the Phase 3 tests solely for the purposes for which they were intended—to identify areas needing improvement, not as a justification for certifying whether the ASPs represent a significant increase in operational effectiveness. However, if the Secretary of DHS intends to consider the results of the Phase 3 tests, along with other test data, in making a certification decision regarding ASPs, then we also recommend that the Secretary (1) direct the Director of DNDO to revise and clarify the Phase 3 test report to more fully disclose and articulate the limitations present in the Phase 3 tests—particularly the limitations associated with making comparisons between detection systems from a small number of test runs-—and (2) clearly state which "relevant insights into important aspects of system performance" from the Phase 3 report are factored into any decision regarding the certification that ASPs demonstrate a significant increase in operational effectiveness. Finally, we further recommend that since there are several phases of additional ASP testing currently ongoing, the Secretary should direct the Director of DNDO take steps to ensure that any limitations associated with ongoing testing are properly disclosed when the results of the current testing are reported.

We provided DHS with a draft of this report for its review and comment. The department stated that it strongly disagreed with our draft report and two of our recommendations, it agreed to take some action on a third recommendation, and offered no comments on a fourth recommendation. In its comments, DHS cites narrative from the Phase 3 report explaining that the Phase 3 tests employed fewer test runs per test so as "to allow for more substantial variation among test cases" rather than "running sufficient number of repetitions... to provide high statistical significance results." Thus, in DHS's view, our assertion that the report does not "fully disclose" the Phase 3 test's limitations concerning the statistical significance of the results is incorrect. Our draft report recognizes DHS's description of how the Phase 3 tests were conducted. Our concern is that although DNDO cited the limited statistical significance of the test results at the outset of the Phase 3 report, it does not clearly state this limitation in expressing the test's findings. For example, as we note in our draft report, the Phase 3 report repeatedly states that the performances of the various ASPs were "statistically indistinguishable" even though DNDO did not perform enough test runs to estimate with a high degree of confidence whether the performances were actually similar. DNDO presents many of its findings as conclusive statements about ASP performance despite the fact that the Phase 3 test design cannot support these findings. DHS had additional comments, which are discussed at the end of this letter.

BACKGROUND

In the summer of 2005, DNDO tested ASPs from 10 vendors to evaluate their performance capabilities and to select the ASPs that warranted further development and possible procurement. In July 2006, DNDO awarded contracts totaling $1.2 billion over fiveyears to three vendors—Raytheon, Canberra, and Thermo.

The Department of Homeland Security Appropriations Act for Fiscal Year 2007 states that "none of the funds appropriated ... shall be obligated for full scale procurement of [ASP] monitors until the Secretary of Homeland Security has certified ... that a significant increase in operational effectiveness will be achieved." Congress enacted a similar requirement in DHS's fiscal year 2008 appropriation. In hopes of obtaining secretarial certification by June 2007, DNDO tested ASPs at several sites, including the Nevada Test Site, the New York Container Terminal, the Pacific Northwest National Laboratory, and five ports of entry. DNDO conducted the tests at NTS in two phases. DNDO stated that the Phase 1 tests, performed in February-March 2007, attempted to estimate the performance capabilities of the ASPs with a high degree of statistical confidence. DNDO intended these tests to support the Secretary's decision on whether to certify the ASPs for the purposes of a full-scale production decision, while the Phase 3 tests were intended to help improve the computer algorithms that the ASPs use to identify the specific radiological or nuclear source inside a container.

On September 18, 2007, we testified that DNDO's Phase 1 tests did not constitute an objective and rigorous assessment of the ASPs' capabilities because, among other things, DNDO conducted preliminary test runs on source materials to be used in the tests, and then allowed the vendors to adjust their ASPs to specifically identify the source materials to be tested. We testified that in our view, DNDO's approach biased the tests in ways that enhanced the apparent performance of the ASPs. We also noted that the tests did not attempt to estimate the limits of ASPs' detection abilities—an important concern to those who will use them such as CBP officers. During that hearing, DNDO's Director stated that, contrary to statements DNDO made in its final Phase 3 test plan, DNDO would use the Phase 3 test results to help support the Secretary's decision on whether to certify the ASPs for full-scale production. Subsequently, DNDO delayed its anticipated date for secretarial certification to the fall of 2008 in order to conduct additional performance tests and field tests during fiscal year 2008.

DNDO'S PHASE 3 TEST REPORT FREQUENTLY OVERLOOKS THE LIMITATIONS ASSOCIATED WITH THE TESTS' SMALL SAMPLE SIZES

Because the limitations of the Phase 3 test results are not properly discussed in the Phase 3 test report, the report does not accurately portray the results from the Phase 3 tests and could be misleading. The purpose of the Phase 3 tests was to identify areas in which the ASPs needed improvement. While some of the Phase 3 report addresses this purpose, much of the report compares the performance of the ASPs with each other or with PVTs and RIIDs during the tests. However, because DNDO performed each test a limited number of times, the data it used to make some of these comparisons provide little information about the actual capabilities of the ASPs. The narrative of the report often presents each test result as a single value, although, because of the limited number of test runs, the results would be more thoroughly and appropriately stated as a range of potential values. In addition, the report's narrative sometimes omits key facts that conflict with DNDO's analysis of the results.

The Phase 3 Test Report Largely Overlooks the Limiting Effects of Performing a Small Number of Tests

The purpose of the Phase 3 tests was to provide data that would help further develop and improve the ASPs' identification software and data to aid in the development of secondary screening operations and procedures. DNDO acknowledged early in the test report that the Phase 3 tests did not involve enough test runs to assess the performance of the ASPs with a high degree of statistical confidence:

> "The primary goals of the testing were to provide information by giving the ASP systems an opportunity to perform against a wider range of radionuclides, shielding, and cargo configurations. To allow for more substantial variation among test cases, the test plan called for a smaller number of trials over a larger range of objects and configurations rather than running sufficient number of repetitions for each individual test case to provide higher statistically significan[t] results." (p. 2)

DNDO also acknowledged early in the Phase 3 report that, given the small number of test runs, it would be difficult to compare the performance of the ASPs with each other:

> In these comparisons [of the performances of different ASP systems], results are often indistinguishable [i.e., not statistically significantly different] because the small sample sizes induce large uncertainties in the estimates of the probabilities being compared [for example: n ≤ 5]."[4] (p.9)

Nonetheless, while some of the Phase 3 report addresses the stated purpose of the tests, the preponderance of the report compares the performance of the ASPs with each other or with PVTs or RIIDs during the tests, as shown in the following examples:

> "For [category of source material] at 2 mph, the ASP system performances are statistically indistinguishable." (p.13)
>
> "For shielded [category of source material], performance for all three systems is statistically indistinguishable with probabilities of correct alarm varying approximately between 0.84 and .0.92." (p.11)

The statements imply that the performances of the ASPs were similar because the results were "statistically indistinguishable." However, given the small number of test runs, it is impossible to determine with a high degree of confidence whether or not the performances were actually similar. Yet the report's text describing specific results rarely qualifies the results by stating that the test was run only a few times or that the results should not be considered conclusive of the ASPs' capabilities.

Similarly, the report's executive summary presents the test results as conclusions about the performance capabilities of the ASPs, PVTs, and RIIDs:

"For the source configurations tested, the ASP systems have equal performance down to the lowest source activity tested." (p. iii)

"The PVT systems display lower performance than the ASP systems for [category] and [category] sources." (p. iv)

"When comparing the ASP systems _ mph identification metric with the _ RIID measurements..., it is observed that the RIID performance is poor compared to the ASP systems." (p. iv)[5]

Report Text Often Omits the Range of Values Surrounding Each Test Result

The Phase 3 test report makes some of its performance comparisons by citing the percentage of correct detections or identifications that an ASP made on a test. For example:

"For bare [category of source material] only, ... [T]he probability of correct identification [for the 3 ASPs] varied between 0.34 and 0.5." (p.14)[6]

However, because each test involved a small number of test runs, these percentages provide little information about the performance capabilities of the ASPs. In fact, because of the small number of test runs, DNDO can only estimate that each ASP can correctly identify the type of source material within a range of values.[7] The fewer the number of test runs, the larger the range. For example, for the ASP that correctly identified the source material 34 percent of the time during the tests, the report text omits the fact that, as shown on an accompanying graph, DNDO can only estimate that the ASP would be able to correctly identify the source between about 10 percent and 65 percent of the time. By stating that the ASP identified the source 34 percent of the time without clarifying that the results came from only a few test runs, the report's text makes the test results seem more precise than they really are. Similarly, for the ASP that correctly identified the source material 50 percent of the time during the tests, the small number of test runs means that DNDO can only estimate that the ASP would be able to correctly identify the source material between about 15 percent and 85 percent of the time. This range is too wide to have much value in determining how well the ASP may perform in the real world. Although these ranges are clearly shown on the report's graphs, they are omitted in the report's descriptions and interpretations of the test results.

Similarly, DNDO's analysis comparing the performances of ASPs and RIIDs fails to consider the uncertainties created by the tests' small sample sizes. The report states that the RIIDs "performance is poor compared to the ASP systems." For example, during the tests, one vendor's ASP correctly identified one type of source material about 50 percent of the time, while the RIIDs correctly identified the same type of source material about 10 percent of the time. However, given the small number of test runs, DNDO cannot be confident that these percentages precisely indicate the performance capabilities of the ASPs and RIIDs. On the basis of the tests, DNDO can only infer that the ASPs' and RIIDs' performance capabilities lie somewhere within a relatively large range of values. As these ranges are illustrated in the report's graphs, it appears that the difference in the performance of the ASPs and RIIDs may not be statistically different for three of the five types of source materials DNDO tested. This does not necessarily mean that the ASPs and RIIDs performed equally well; rather, DNDO did not conduct each test enough times to determine that the superior performance of the ASPs over the RIIDs reflects the capabilities of the ASPs rather than mere chance.

DNDO's Phase 3 Test Report Omits Important Details that Affect the Interpretation of the Test Results

In a few instances, the report omits important details concerning DNDO's interpretation of the results. For example, DNDO seems to assert in the report's executive summary that the ASPs are as good as the PVTs at detecting radiological or nuclear source materials:

> "The ASP systems demonstrated detection limits equal to or better than those of any of the PVT systems as configured during testing." (p.iii)

However, the report's executive summary fails to note that because DNDO used only one type of source material, the results are largely specific to that particular source material and would not necessarily apply to other specific source materials. In fact, for other types of source material, the report shows several instances in which the PVTs were apparently able to detect other types of source materials better than the ASPs. Moreover, other Phase 3 tests showed that simply moving the source material referred to in the above quote to another place in the container affected the relative performances of the ASPs and PVTs.

Similarly, in reporting how well the ASPs performed when the radiation from the source material was partially blocked by a shielding material, DNDO stated:

> "the ASP systems have the ability to identify sources when placed inside almost all but the thickest shielding configuration tested." (p.iv)

Again, however, DNDO fails to note in its report that, as it explained in its Phase 3 test plan, all the shielding used in the Phase 3 tests represented "light shielding." The report also fails to state how many specific sources the ASPs could correctly identify or how frequently the ASPs could identify them.

Phase 3 Test Results Provide Little Evidence as To Whether ASPs Represent an Improvement Over Currently Deployed Technology

In our view, it is not appropriate to use the Phase 3 test report in determining whether the ASPs represent a significant improvement over currently deployed radiation equipment because the limited number of test runs does not support many of the comparisons of ASP performance made in the Phase 3 report. As noted, DNDO's use of a small number of runs for each test means that DNDO can only be certain that the ASP can correctly identify or detect a source material over a broad range of possible values rather than at a specific rate. This is true even if the ASP was successful every time a test was conducted. For example, as noted in the Phase 3 test report, if the ASP correctly identified a source material 100 percent of the time, but the test was run only five times, the most DNDO can estimate is that the ASP should be able to correctly identify the source no worse than about 60 percent of the time.

The Phase 3 test results do not help to determine an ASP's "true" level of performance because DNDO did not design the tests to assess ASP performance with a high degree of statistical confidence. In the Phase 3 test plan, DNDO was very clear that it had intended the tests to help develop a conduct of operations for secondary screenings and to cover a larger array of source materials and test scenarios than were conducted in the Phase 1 tests.

DNDO also originally stated that the Phase 3 tests would not be used for secretarial certification that the ASPs represented a "significant operational improvement" over currently deployed radiation detection equipment. DNDO stated that it had designed the Phase 1 tests to "evaluate the current state of performance of the ASP...systems." However, prior to releasing the Phase 3 report, DNDO changed its position, stating in the final Phase 3 test report that the test results are relevant to secretarial certification:

> "The Phase 3 test campaign was not originally intended to support the Secretarial Certification of the ASP systems. However, the test results provide relevant insights into important aspects of system performance and should be taken into consideration by the Secretary of Homeland Security in making his (ASP procurement) decision." (p.iii)

It is important to note that DNDO does not elaborate in the test report as to what the "relevant insights" are or how they relate to Secretarial certification. DNDO also does not explain why those insights would be relevant considering that, as stated in the Phase 3 test plan, the results from the tests lack a high degree of statistical significance. Finally, it should be noted that when the Director of DNDO testified in September 2007 that the Phase 3 test results would help inform the Secretary's recommendation, he also acknowledged that the Phase 3 test report had not yet been prepared.

SNL's Special Tests Were Designed to Improve ASP Performance

The special tests were performed by experts from Sandia National Laboratories who were not part of the Phase 1 or Phase 3 tests. The special tests were designed to examine potential vulnerabilities associated with either the ASPs or the Phase 1 or Phase 3 test plan and vulnerabilities in DNDO's test processes. Conducting this type of test would allow the ASP vendors the opportunity to make improvements to their systems in order to address weaknesses revealed during the special tests. Like the Phase 3 tests, the special tests used a small number of runs for each testing scenario. Because of the small number of runs, the test results do not support estimating the probability of detection or identification with a high confidence level making it difficult to use the results of the special tests to support a certification decision by the Secretary of DHS. On this point, the special test report acknowledges that "the special tests were never intended to demonstrate conformity of the [ASP] systems against specific performance requirements."

From the special tests, SNL concluded

1. "Areas for software and hardware improvement have been identified based on system performance issues observed for the versions of the ASP hardware and software tested at the NTS during Winter 2007."
2. "For the data made available to us, the reported results ... are consistent with the underlying collected data—indicating that the DNDO ASP system assessment was not biased."
3. "Recommendations to improve the testing rigor have been made...(noting that) their implementation must be balanced against other test campaign impacts (such as) cost, schedule, availability of resources, etc.," and
4. "Based on our limited tests we observed no data suggesting that the ASP system performance was inappropriately manipulated by either the vendors or the test team."

Overall, the special test report appears to accurately describe the purpose, limitations, and results of the special tests. In our view, DNDO should consider SNL's views as it proceeds with additional ASP testing in 2008. It is important to note, however, in Sandia's conclusions that the "ASP system assessment [in 2007] was not biased" and that it "observed no data suggesting that the ASP system performance was inappropriately manipulated," Sandia is referring to the data derived from ASP tests. However, SNL does not comment on the biased testing methods we identified during the Phase 1 ASP tests at the Nevada Test Site in 2007. Specifically, when we stated in September 2007 that DNDO's Phase 1 tests were biased, we were referring to DNDO's test methods which (1) used the same test sources and shielding materials during preliminary runs as were used during the actual tests and (2) did not use standard CBP operating procedures in testing the RIIDs.

Conclusions

Preventing the material for a nuclear weapon or a radiological dispersal device from being smuggled into the United States remains a key national security priority. Testing radiation detection equipment to understand its capabilities and limitations is an important part of preventing nuclear smuggling. The Phase 3 and special tests were part of DNDO's 2007 effort to test ASPs in order to identify areas for further development to these devices. The Phase 3 test results are relevant to DNDO's original objective for the Phase 3 tests—to identify areas in which the ASPs needed improvement. However, because of the limitations of the tests, DNDO should not be using the test results as indicators of the overall performance capabilities of the ASPs. Moreover, in the Phase 3 report, DNDO presented and analyzed the test results without fully disclosing key limitations of the tests, which is not consistent with basic principles of statistics and data analysis. Because of this, many of the report's presentations and comparisons of performance among ASPs and between ASPs and PVTs are not well supported and are potentially misleading. Regarding the special tests, SNL notes in its test report that it designed the tests to identify areas where the ASPs need to improve—not to measure the ASPs performance against requirements. Overall, because of the limitations discussed in this report, it is our view that neither the Phase 3 tests nor the special tests should serve as a basis for the Secretary of DHS whether the ASPs represent "a significant increase in operational effectiveness" over current radiation detection equipment.

Recommendations for Executive Action

To ensure that the limitations of the Phase 3 test results, and future ASP test results, are clearly understood, we are making the following four recommendations.

We recommend that the Secretary of DHS use the results of the Phase 3 tests solely for the purposes for which they were intended—to identify areas needing improvement, not as a justification for certifying whether the ASPs warrant full-scale production.

However, if the Secretary of DHS intends to consider the results of the Phase 3 tests, along with other test data information, in making a certification decision regarding ASPs, then we recommend that the Secretary take the following actions:

- Direct the Director of DNDO to revise and clarify the Phase 3 test report to more fully disclose and articulate the limitations present in the Phase 3 tests—particularly the limitations associated with making comparisons between detection systems from a small number of test runs.
- Clearly state which "relevant insights into important aspects of system performance" from the Phase 3 report are factored into any decision regarding the certification that ASPs demonstrate a significant increase in operational effectiveness.

Finally, we further recommend that since there are several phases of additional ASP testing currently ongoing, the Secretary should direct the Director of DNDO take steps to ensure that any limitations associated with ongoing testing are properly disclosed when the results of the current testing are reported.

Comments from the Department of Homeland Security and Our Evaluation

We provided DHS with a draft of this report for its review and comment. Its written comments are presented in appendix I. The department stated that it strongly disagreed with our draft report and two of our report's recommendations. DHS agreed to take some action on a third recommendation and offered no comments on a fourth recommendation. The department stated several reasons for its disagreement. First, DHS cites narrative from the Phase 3 report explaining that the Phase 3 tests employed fewer test runs per test so as "to allow for more substantial variation among test cases" rather than "running sufficient number of repetitions … to provide high statistical significance results." Thus, in DHS's view, our assertion that the report does not "fully disclose" the Phase 3 tests' limitations concerning the statistical significance of the results is incorrect. Our draft report recognizes DHS's description of how the Phase 3 tests were conducted. Our concern is that although DNDO cited the limited statistical significance of the test results at the outset of the Phase 3 report, DNDO's findings do not reflect this limitation. For example, as we note in our draft report, the Phase 3 report repeatedly states that the performances of the various ASPs were "statistically indistinguishable" even though DNDO did not perform enough test runs to estimate with a high degree of confidence whether the performances were actually similar.

DNDO presents many of its findings as conclusive statements about ASP performance despite the fact that the Phase 3 test design cannot support these findings.

Second, the department commented that the Phase 3 test report clearly and succinctly stated another limitation of the test methodology—specifically, that the tests were not designed to be a precise indicator of ASP performance. In the department's view, noting this limitation throughout the Phase 3 report would have been unwieldy. We did not expect DNDO to repeat this limitation throughout the report. However, as suggested in our report, the Phase 3 report should accurately reflect the test results without portraying the results as being more precise than they really are. Using an example from the Phase 3 report, if DNDO notes that an ASP successfully identified a specific source material 34 percent of the time during the tests, it should also indicate that, given the small number of test runs, DNDO can only estimate that the ASP would be able to correctly identify the specific source material between 10 and 65 percent of the time. However, no such discussion of the wide range of potential results is included in the report's narrative. In our view, presenting the test results without sufficient narrative about the tests' limitations is potentially misleading.

Third, the department stated that although the Phase 3 tests were not intended to support the DHS Secretary's certification decision, DHS decided that it needed to consider all available test results in making this decision. DHS further commented that not doing so would subject it to criticism of "cherry-picking" the results. In response, although we acknowledge the need to consider all available test results, we believe they should be considered in their appropriate context, and that test results do not all carry the same weight. In our view, test results with a high degree of statistical significance (i.e., unlikely to be the result of chance) should be considered a better indicator of ASP performance than those with a lower level of statistical significance. Because the Phase 3 tests involved only 1-10 runs per test, very few of the results can be generalized as reliable estimates of how the ASPs perform and thus potentially provide questionable evidence for the certification process. We also note that, in its comments, DHS did not address what Phase 3 results or important insights it considered to be relevant to Secretarial certification.

Fourth, DHS comments that our draft report failed "to acknowledge the depth and breadth of the ASP test campaign, which is by far the most comprehensive test campaign ever conducted on radiation detection equipment." However, our report describes previous ASP testing and some of our prior findings about that testing, and notes that ASP testing continues in 2008. More importantly, the extent of testing is not the issue at hand. In our view, regardless of how many tests are performed, the tests must employ sound, unbiased methodologies and DNDO should draw and present conclusions from the test results in ways that accurately and fully reflect the data and disclose their limitations.

DHS stated that it disagreed with our recommendations to (1) use the Phase 3 test's to identify areas needing improvement and not as a basis for certification and (2) revise and clarify the Phase 3 report to reflect the limitations in the tests' methodology and results. It did not offer comments on our recommendation that the Secretary clearly state what relevant insights from the Phase 3 report are factored into any certification decision. We continue to believe that the Phase 3 tests should be used only for the intended purpose stated in its test plan—to improve the software of ASPs. We would also note that our draft report recommends that DNDO revise and clarify the Phase 3 test report only if it includes Phase 3 test results among the data that will be presented to the Secretary prior to his decision on certification. If DNDO chooses to use the Phase 3 test results for certification, we believe it is important that DNDO explain what test results are relevant to certification and why the value of those results are not mitigated by the limitations associated with the Phase 3 tests' small sample sizes.

In response to our last recommendation, the department stated that it has taken and will continue to take steps to ensure that it properly discloses any limitations associated with ongoing testing as it moves toward secretarial certification of the ASPs.

As agreed with your offices, unless you publicly announce the contents of this report, we plan no further distribution until 30 days from the report date. At that time, we will send copies of this report to the Secretary of DHS and interested congressional committees. We will also make copies available to others upon request. In addition, this report will be available at no charge on the GAO Web site at http://www.gao.gov.

If you or your staffs have any questions about this report, please contact me at (202) 512-3841 or aloisee@gao.gov. Contact points for our Offices of Congressional Relations and Public Affairs may be found on the last page of this report. GAO staff that made major contributions to this report are listed in appendix II.

Gene Aloise
Director,
Natural Resources and Environment

Appendix I. Comments from the Department of Homeland Security

U.S. Department of Homeland Security
Washington, DC 20528

Homeland Security

August 29, 2008

Mr. Gene Aloise
Director, Natural Resources and Environment
Government Accountability Office
441 G Street NW
Washington, DC 20548

Dear Mr. Aloise:

Re: Draft Report GAO-08-979, Combating Nuclear Smuggling: DHS's Phase 3 Test Report on Advanced Portal Monitors Does Not Fully Disclose the Limitations of the Test Results.

The Department of Homeland Security strongly disagrees with the GAO Draft Report. The title of the GAO report "DHS's Phase 3 Test Report on Advanced Portal Monitors Does Not Fully Disclose the Limitations of the Test Results" is misleading and not substantiated by the body of the GAO report. First, the Test Purpose section of the test report clearly states that "to allow for more substantial variation among test cases, the test plan called for a smaller number of trials over a larger range of objects and configurations rather than running sufficient number of repetitions for each individual test case to provide high statistical significance results."[1] This statement applies to the entire test report; therefore to say that DHS did not "fully disclose" limitations that could be "potentially misleading" is false. Repeating the same statement throughout the test report would be unnecessary and tedious.

Second, DHS clearly acknowledges that the main purpose of the Phase 3 test was to provide information to improve the system algorithms.[2] The ASP Phase 3 test was designed to identify areas for ASP development and improvement. The test was not intended to be a precise indicator of ASP performance, nor does the test report ever claim to draw such conclusions. The underlying test design purposely involved test cases that were different from those against which ASP performance is to be measured. The test report clearly and succinctly stated the limitations associated with the test methodology and analysis approach. Again, given the amount of

[1] "Test Report in Support of Advanced Spectroscopic Portal (ASP) Systems Development at the Nevada Test Site (NTS)", Section 1.2 Test Purpose, Page 2

[2] "Test Report in Support of Advanced Spectroscopic Portal (ASP) Systems Development at the Nevada Test Site (NTS)", Section 1.2 Test Purpose, Page 2 and Section 2 Test Results, Page 4

material and density of the test report, it would have been unwieldy to repeat caveats throughout the report.

Third, even though the Phase 3 report was not designed to support the certification decision or to substantiate performance of the systems against the defined threat basis, DHS quickly recognized that it needed to consider all available test results in making a certification decision. For that reason, the Phase 3 Test report clearly states "Although the Phase 3 test campaign was not originally intended to support the secretarial certification of the ASP systems, the test results provide relevant insights into important aspects of the system performance and should be taken into consideration by the Secretary of Homeland Security."[3] Indeed, had we not done so, DHS would now be accused of "cherry-picking" test results, or worse, ignoring data.

Once again, by reporting on ASP testing in a piece-meal fashion, GAO fails to acknowledge the depth and breadth of the ASP test campaign, which is by far the most comprehensive test campaign ever conducted on radiation detection equipment. The ASP Phase 3 test is but one in a series of carefully-designed tests conducted over a period of years in the path to secretarial certification and full-rate production. To date, these campaigns have included a developmental set of testing for the engineering developmental models in Winter 2007 and performance testing at the Nevada Test Site (NTS), deployment readiness testing at the Pacific Northwest National Laboratory (PNNL), operational testing at the New York Container Terminal (NYCT), and field validations at multiple POEs conducted in Summer 2008.

In conjunction with the Department of Energy (DOE), DNDO is also conducting Threat Injection Studies for ASP systems. These threat injection studies will examine the limits of performance of ASP in order to guide the setting of thresholds and concept of operations (CONOPS) and also compare ASP and polyvinyl toluene (PVT) performance. To date, DNDO has developed, integrated, and validated a set of tools to perform the injection studies using a standard data format. Data collection has also been underway to create threat-representative signatures that can be injected into stream-of-commerce data. In preparation for the studies, DNDO has worked to collect data to validate injection methodology and prepare data set of approximately 8000 validated stream-of-commerce data files into which threat signatures will be injected and also create threat-representative signatures based on the collected data.

Additional 2008 ASP testing is currently underway, and includes: system qualification testing (SQT) to verify compliance with the ASP performance specification; integration testing at PNNL to verify that ASP performance remains sound when it is integrated into the POE architecture; performance testing at NTS to validate the detection and identification capabilities of ASP systems against special nuclear materials (SNM) and materials for radiological dispersal devices (RDD); and operational test and evaluation activities to validate operational performance of the system at POEs. The successful completion of these steps will provide data for the Secretary's Certification decision. DNDO will use a combination of cost-benefit analyses as well as demonstrated performance metrics to assist in the Secretary's certification decision. Part of the certification process will involve working with the National Academy of Sciences, to review DNDO test plans and procedures, as required in the FY 2008 Consolidated Appropriations Act.

---

[3] "Test Report in Support of Advanced Spectroscopic Portal (ASP) Systems Development at the Nevada Test Site (NTS)", Section 1.2 Test Purpose, Page 2 and Executive Summary, Page iii

DHS will also use the test results, along with other information, to seek approval of the DHS Investment Review Board (IRB) prior to proceeding to full-scale production and deployment of ASP systems at POEs.

Based on the above, DHS believes that the Phase 3 Test Report more than adequately discloses the purpose and limitations associated with the Phase 3 test results, and therefore, disagrees with the GAO's recommendation that the Phase 3 report needs to be revised and clarified. DHS also believes that all data needs to be considered for the secretarial certification decision, and therefore, disagrees with the GAO's recommendation that the Phase 3 data not be considered. It is neither prudent nor scientifically justifiable to pick the data one chooses to use for making a decision. While some data may carry more weight than others (particularly the most recent data for the newest software version), no data should be ignored. DHS has taken, and will continue to take, steps to ensure that any limitations associated with ongoing testing are properly disclosed in the path to certification.

Thank you for the opportunity to review and provide comments to the draft report.

Sincerely,

Jerald E. Levine

Jerald E. Levine
Director, Departmental GAO/OIG Liaison Office

End Notes

[1] According to DNDO, Phase 2 was its completion of the test report for the Phase 1 tests.

[2] Equipment testing involves repeating a single test multiple times to estimate how often a device performs its function correctly. The set of tests is referred to as a sample. Each test in the sample is considered a random trial and therefore the estimates derived from the sample are subject to random variations; in other words, if the series of tests were repeated then each sample could yield different estimates. Because of these random variations, estimates from samples are often presented as a range of possible values called the 95 percent confidence interval. This is the range of values that will contain the true probability of performance in 95 percent of the samples that we might select. In general, when the sample size (number of tests) is larger, the range of possible values is smaller, allowing more precise estimates of the likely performance of the machine.

[3] GAO, *Combating Nuclear Smuggling, Additional Actions Needed to Ensure Adequate Testing of Next Generation Radiation Detection Equipment*, GAO-07-1247T (Washington, D.C.: Sept. 18, 2007).

[4] The report does not present significance-of-differences tests with its analyses.

[5] DNDO's Phase 3 report is classified. Because of this, some of the quotes from the report are missing information in order to protect sensitive information.

[6] DNDO analyzed this series of tests by source category (medical, industrial, or special nuclear material) rather than by specific source material or isotope. For ease of discussion, we refer to each category as a source material.

[7] Unless stated otherwise, the range of values represents the confidence intervals surrounding the point estimate at the 95 percent level.

In: Detection of Nuclear Weapons and Materials
Editor: Pietro Egidi
ISBN: 978-1-60741-512-1

Chapter 2

DETECTION OF NUCLEAR WEAPONS AND MATERIALS: SCIENCE, TECHNOLOGIES, OBSERVATIONS

Jonathan Medalia

SUMMARY

Detection of nuclear weapons and special nuclear material (SNM, certain types of uranium and plutonium) is crucial to thwarting nuclear proliferation and terrorism and to securing weapons and materials worldwide. Congress has funded a portfolio of detection R&D and acquisition programs, and has mandated inspection at foreign ports of all U.S.-bound cargo containers using two types of detection equipment.

Nuclear weapons contain SNM, which produces unique or suspect signatures that can be detected. It emits radiation, notably gamma rays (high-energy photons) and neutrons. SNM is very dense, so it produces a bright image on a radiograph (a picture like a medical x-ray) when x-rays or gamma rays are beamed through a container in which it is hidden. Using lead or other shielding to attenuate gamma rays would make this image larger. Nuclear weapons produce detectable signatures, such as radiation generated by or a noticeable image on a radiograph. Other detection techniques are also available.

Nine technologies illustrate the detection portfolio: (1) A new scintillator material to improve detector performance and lower cost. (2) GADRAS, an algorithm to determine the materials in a container by analyzing gamma-ray spectra. If materials are the "eyes and ears" of detectors, algorithms are the "brains." (3) A project to simulate large numbers of experiments in order to improve detection system performance. (4, 5) Two Cargo Advanced Automated Radiography Systems (CAARS) to detect high-density material based on the principle that it becomes less transparent to photons of higher energy, unlike other material. (6) A third CAARS to detect material with high atomic number (Z, number of protons in an atom's nucleus) based on the principle that Z affects how material scatters photons. (7) A system to generate a 3-D image of the contents of a container based on the principle that Z and density strongly affect the degree to which muons (a type of subatomic particle) scatter.

(8) Nuclear resonance fluorescence imaging to identify materials based on the spectrum of gamma rays a nucleus emits when struck by photons of a specific energy. (9) The Photonuclear Inspection and Threat Assessment System to detect SNM up to 1 km away, unlike other systems that operate at very close range. It would beam high-energy photons at distant targets to stimulate fission in SNM, producing characteristic signatures that may be detected.

This analysis leads to several observations for Congress. Some detection technology is advancing faster than many have expected. It is easier and less costly to accelerate a program in R&D than in production. "Concept of operations" is crucial to detection system effectiveness. Congress may wish to address gaps and synergisms in the technology portfolio. Congress need not depend solely on procedures developed by executive agencies to test detection technologies, but may specify tests an agency is to conduct. Ongoing improvement in detection capabilities produces uncertainties for terrorists that will increase over time, adding deterrence beyond that of the capabilities themselves. This report will be updated occasionally.

Congress has been interested in detecting nuclear weapons and the special materials needed to make them for many years, especially since 9/11. Nuclear detection has many applications for countering nuclear terrorism and nuclear proliferation, such as securing nuclear weapons and materials in U.S., Russian, and other nuclear facilities, tracking materials at border crossings and choke points, screening maritime cargo containers, and examining actual or suspect nuclear sites.

The United States currently uses several types of nuclear detection equipment. All have significant shortcomings. Some work only at very short range; some cannot identify the material emitting radiation, which can lead to false alarms and interrupt commerce; some depend on operator skill, and may be defeated by a clever smuggler or a sleepy operator; and some are easily defeated by shielding. In an effort to overcome such problems, Congress has funded a pipeline of advanced-technology research, development, and acquisition.

This report seeks to help Congress understand this technology. It discusses the science of detecting nuclear weapons and materials, describes nine advanced U.S. technologies selected to illustrate the range of projects in the pipeline, and offers observations for Congress. The report does not compare technologies.[1] The inclusion of the nine technologies should not be taken to mean that CRS judges them to be better than the hundreds of others not considered here. The report does not discuss the controversial Advanced Spectroscopic Portal because a detailed discussion of it could draw attention from the other technologies considered here.

The scope of this report excludes the organization of the government for dealing with nuclear detection, the role of intelligence and law enforcement in detecting terrorist nuclear weapons, detection of radiological dispersal devices (such as "dirty bombs"), the role of nuclear forensics in attributing an attack to its perpetrator, response to a nuclear attack, and the architecture of a national nuclear detection system.[2] Nor does it discuss possible means by which terrorists might acquire a bomb, or whether they could make a bomb on their own. Much has been written on these topics.[3] While many who are concerned with nuclear detection focus on thwarting nuclear terrorism, this report focuses on technology per se. It avoids extensive discussion of means to counter detection to avoid classified information.

Nuclear detection technology has a dual role in thwarting a terrorist nuclear attack—deterrence and defense. Deterrence means dissuasion from an action by threat of unacceptable consequences. Terrorists may be deterred from a nuclear strike by one of the few consequences unacceptable to them: failure. Detection systems would raise that risk. These systems could also make a terrorist nuclear strike too complex to succeed. But other factors would also have these effects: the difficulty of fabricating a bomb, the chance that law enforcement or intelligence would detect efforts to obtain a bomb, the possible inability to detonate a purloined bomb, and the risk that scientists recruited for the plot would defect. Such risks would disappear, however, if terrorists were given a bomb and operating instructions. They would then only need to mount a smuggling operation. In that case, the role of nuclear detection systems would change: they would become the main defense.

1. Nuclear Weapons and Materials: Signatures and Detection

As background for understanding the detection technologies in Chapter 2, this chapter outlines nuclear detection science. The **Appendix** provides more detail.

What Is to Be Detected?

Detection focuses on weapons and the nuclear materials that fuel them. Weapons can be small. In the Cold War, the United States built high-yield weapons several feet long, atomic demolition munitions that a soldier could carry, and nuclear artillery shells. A terrorist-made weapon would probably be larger.

Nuclear weapons require fissile material, atoms of which can fission (split) when struck by fast or slow neutrons[4]; pieces of this material can support a nuclear chain reaction. The fissile materials used in nuclear weapons are uranium, isotope 235 (U-235), and plutonium, isotope 239 (Pu-239). The Atomic Energy Act of 1954 designates them as "special nuclear material" (SNM).[5] Uranium in nature is 99.3% U-238 and 0.7% U-235; U-235 must be concentrated, or "enriched." Uranium enriched to 20% U-235 is termed highly enriched uranium, or HEU, but nuclear weapons typically use uranium enriched to 90% or so. In this report, HEU refers to this weapons-grade enrichment level. Weapons-grade plutonium, or WGPu, is also a mix of isotopes, at least 93% Pu-239. According to one account by five nuclear weapon scientists from Los Alamos National Laboratory, it would take 26 kg of HEU or 5 kg of WGPu to fuel a bomb,[6] amounts that would fit into cubes 11 cm or 6 cm, respectively, on a side.

Photons 101

Nuclear detection makes extensive use of photons, packets of energy with no rest mass and no electrical charge. Electromagnetic radiation consists of photons, and may be measured as wavelength, frequency, or energy; for consistency, this report uses only energy, expressed in units of electron volts (eV). Levels of energy commonly used in nuclear detection are thousands or millions of electron volts, keV and MeV, respectively. The electromagnetic spectrum ranges from radio waves (some of which have photon energies of 10^{-9} eV), through visible light (a few eV), to higher-energy x-rays (10 keV and up) and gamma rays (mostly 100 keV and up).

X-ray photons and gamma-ray photons of the same energy are identical. However, they are generated in different ways. Gamma rays originate in processes in an atom's nucleus. Each radioactive isotope that emits gamma rays does so in a unique energy spectrum, as in **Figure 1**, which is the only way to identify an isotope outside a laboratory. A detector with a form of "identify" or "spectrum" in its name, such as Advanced Spectroscopic Portal or radioactive isotope identification device, identifies isotopes by their gamma-ray spectra. X-rays originate in interactions with an atom's electrons. Many detection systems use x-ray beams, which can have higher energies than gamma rays and thus are more penetrating. X-ray beams are often generated through the *bremsstrahlung* process, German for "braking radiation," which works as follows. An accelerator creates a magnetic field that accelerates charged particles, such as electrons, which slam into a target of heavy metal. When they slow or change direction as a result of interactions with atoms, they release energy as x-rays whose energy levels are distributed from near zero to the energy of the electron beam. They do not exhibit spectral peaks like gamma rays.

What Signatures show the presence of nuclear weapons and SNM?

For purposes of this report, a signature is a property by which a substance (in particular, SNM) may be detected or identified. This section presents several signatures. The Appendix and Chapter 2 discuss others.

Atomic number and density

Atomic number, abbreviated "Z," is the number of protons in an atom's nucleus. It is a property of individual atoms. In contrast, density is a bulk property, expressed as mass per unit volume. In general, the densest materials are those of high Z. These properties may be used to detect uranium and plutonium. Uranium is the densest and highest-Z element found in nature (other than in trace quantities); plutonium has a slightly higher Z (94), and its density varies from slightly more to slightly less than that of uranium. Some detection methods discussed in Chapter 2, such as effective Z, make use of Z, and some, such as radiography and muon tomography, make use of Z and density combined.

Opacity to photons

An object's opacity to a photon beam depends on its Z and density, the amount of material in the path of the beam, and the energy of the photons. Gamma rays and x-rays can penetrate more matter than can lower-energy photons, but dense, high-Z material absorbs or scatters them. Thus a way to detect an object, such as a bomb, in a container is to beam in x-rays or gamma rays to create a radiograph (an opacity map) like a medical x-ray.

Radioactivity

Radioactive atoms are unstable and give off various types of radiation; the types of use for nuclear detection are gamma rays and neutrons.

Gamma rays. Gamma-ray spectra are well characterized for each isotope. **Figure 1** and **Figure 2** show spectra for HEU and WGPu. Each point on the spectrum shows the number of photons emitted (vertical axis) at each energy level (horizontal axis). Background gamma radiation is ubiquitous. Since many materials, including SNM, emit gamma radiation, elevated levels of gamma radiation may or may not indicate the presence of SNM, but careful analysis of the total gamma-ray spectrum, as discussed in Chapter 2 under GADRAS, may reveal the presence of SNM.

Neutrons. Atoms of some heavy elements fission. Of the naturally-occurring elements, only U238 spontaneously fissions at an appreciable rate. Fission releases neutrons. U-235 emits few neutrons, but because U-238, the other main component of HEU, emits some, 1 kg of HEU emits 3 neutrons per second,[7] so it provides a weak signature. Plutonium emits on the order of 60,000 neutrons per kg per second depending on the mix of plutonium isotopes.[8] Unlike gamma rays, neutrons do not have a characteristic energy spectrum by which an isotope can be identified.

How Does Detection Work?

How are signatures gathered, processed, and used?

Detection involves using detector elements to obtain data, converting data to usable information through algorithms, and acting on that information through concept of operations, or CONOPS.

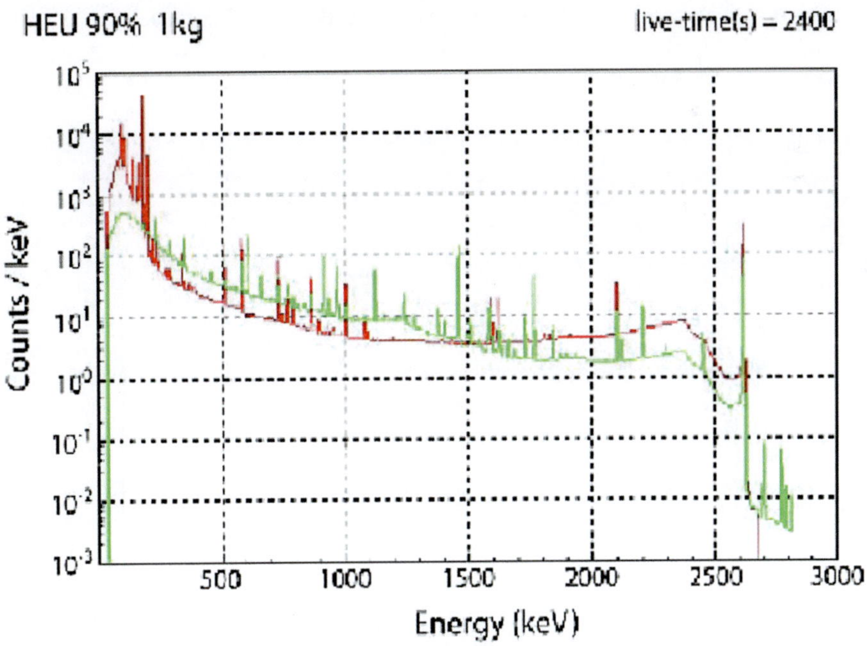

Source: Lawrence Livermore National Laboratory.

Notes: Spectra was taken with a high-purity germanium detector at a distance of 1 m and computed with the GADRAS algorithm, discussed below. Uranium is unshielded, background is dirt. The uranium-235 line starts as the upper line (in red) at far left.

Figure 1. Gamma-Ray Spectra: 90% Uranium-235 vs. Background

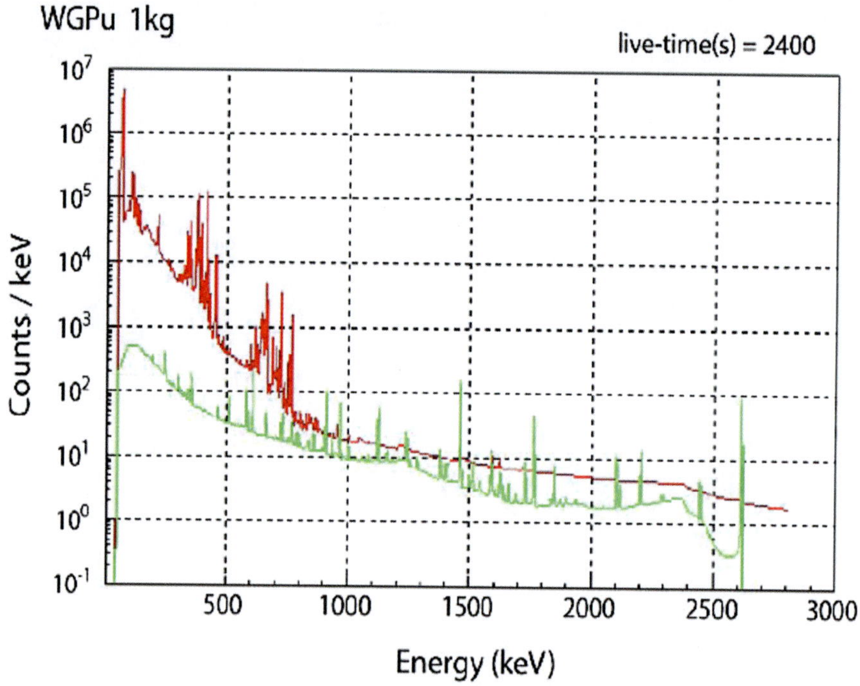

Source: Lawrence Livermore National Laboratory.

Notes: Spectra were taken with a high-purity germanium detector at a distance of 1 m and computer with GADRAS. Plutonium is unshielded, background is dirt. The plutonium line starts as the upper line (in red) at the far left.

Figure 2. Gamma-Ray Spectra: Weapons-Grade Plutonium vs. Background

Detectors, algorithms, and CONOPS are the eyes and ears, brains, and hands of nuclear detection: effective detection requires all three.

Since photons and neutrons have no electrical charge, their energy is converted to electrical pulses that can be measured. This is the task of detectors, discussed next. The pulses are fed to algorithms. An algorithm, such as a computer program, is a finite set of logical steps for solving a problem. For nuclear detection, an algorithm must process data into usable information fast enough to be of use to an operator. It receives data from a detector's hardware, such as pulses representing the time and energy of each photon arriving at the detector. It converts the pulses to a format that a user can understand, such as displaying a gamma ray spectrum or flashing "alarm." Every detector uses one or more algorithms. Improvements to algorithms can contribute as much as hardware improvements to detector capability.

CONOPS may be divided into two parts. One specifies how a detection unit is to be operated to obtain data. Elements include: How many containers must the unit scan per hour? How close would a detector be to a container? Shall the unit screen cargo in a single pass, or shall it be used for primary screening, with suspicious cargo sent for a more detailed secondary screening? A second part details how the data are to be used. Elements include: What happens if the equipment detects a possible threat? Which alarms are to be resolved on-site and which are to be referred to off-site experts? Under what circumstances would a port or border crossing be closed? More generally, how is the flow of data managed, in both directions?[9] What types of intelligence information do inspectors receive, and how do data from detection systems flow to federal, state, and local officials for analysis or action? While this report does not focus on CONOPS because it is not a technology, it is an essential part of nuclear detection.

Principles of detection

Detectors require a signal-to-noise ratio high enough to permit detection. That is, they must extract the true signal (such as a gamma-ray spectrum) from noise (such as spurious signals caused by background radiation). Two concepts are central to gamma-ray detector sensitivity: detection efficiency and spectral resolution. Efficiency refers to the amount of signal a detector records. Radiation intensity (e.g., number of photons per unit of area) diminishes with distance. Since a lump of SNM emits radiation in all directions, using a detector that is larger, or that is closer to the SNM, increases the fraction of radiation from the source that impinges on the detector and thereby increases efficiency. Another aspect is the fraction of the radiation striking the detector that creates a detectable signal. A more efficient detector collects data faster, reducing the time to screen a cargo container.

Spectral resolution refers to the sharpness of peaks in a gamma-ray spectrum. A perfect detector would record a spectrum as vertical "needles" because each radioactive isotope releases gamma rays only at specific energies. Since detectors are not perfect, each energy peak is recorded as a bell curve. The narrower the curve, the more useful the data. Polyvinyl toluene (PVT), a plastic used in radiation detectors that can be fielded in large sheets at low cost, is efficient but has poor resolution. It can detect radiation, but peaks from gamma rays of different energies blur together, which can make it impossible to identify an isotope. **Figure 3** shows the spectra of 90% U-235 and background radiation as recorded by a PVT detector. In contrast, high-purity germanium (HPGe) produces sharp peaks, permitting clear identification of specific isotopes. These detectors are expensive, heavy, have a small detector area, and must be cooled to extremely low temperatures with liquid nitrogen or a mechanical system, making them less than ideal for use in the field. **Figure 4** shows the spectrum of Pu-239 as recorded by detectors with better resolution than PVT.

Sensitivity can be improved. (1) One type of detector is cadmium-zinc-telluride (CZT) crystals. Better crystals and better ways to overcome their limitations have improved sensitivity. The peak on the right of each spectrum in **Figure 5** shows the cesium-137 spectrum taken with CZT detectors that, for the years indicated, were at the high end of sensitivity. (2) Detectors build radiography images or gamma-ray spectra over time. With more time, a detector can collect more data, in the form of gamma rays or neutrons. More data improve separation of signal from noise and reduce false alarms. Longer scan times improve accuracy but impede the flow of commerce, costing money, so a balance is sought between these two opposed goals. (3) Increasing the spatial resolution of a detector improves sensitivity:

> This is easily demonstrated in the example of a shielded versus unshielded radiation detector. Unshielded detectors are sensitive to radiation impinging on it in all directions, which is characteristic of the nature of naturally-occurring background radiation. By adding shielding, a detector's field-of-view can be controlled, and background radiation levels reduced, increasing the signal-to-noise ratio for the detector in the direction the detector is aimed.[10]

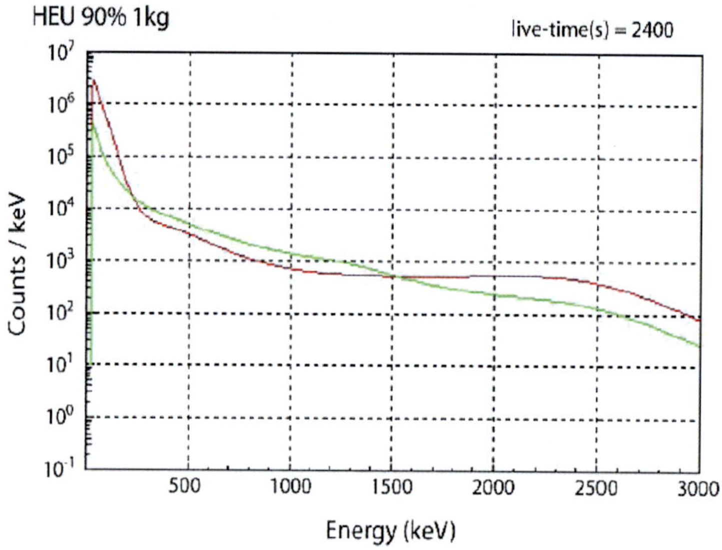

Source: Lawrence Livermore National Laboratory.

Notes: Spectra were taken with a PVT detector at a distance of 1 m and computed with GADRAS. Uranium is unshielded, background is dirt. The uranium-235 line starts as the upper line (in red) at far left.

Figure 3. Gamma-Ray Spectra: 90% Uranium-235 vs. Background, Taken with a PVT Detector

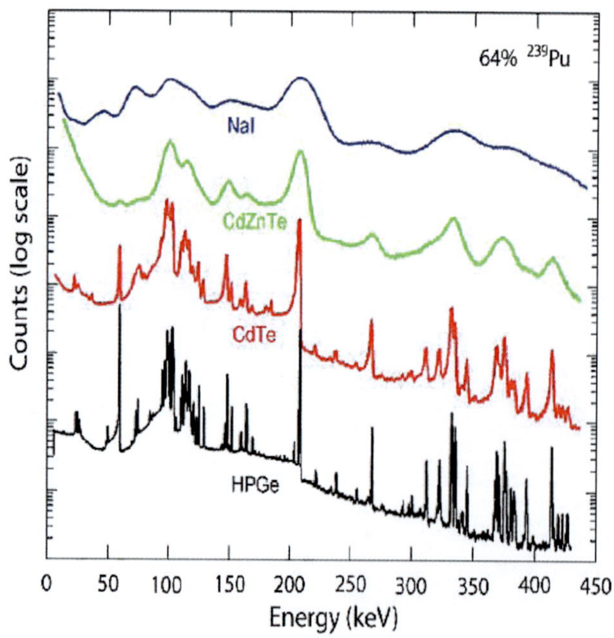

Source: Los Alamos National Laboratory.

Notes: These are spectra of 64% plutonium-239. They were taken using different detector materials to show differences in resolution: from top to bottom, sodium iodide mixed ("doped") with thallium; cadmium-zinctelluride; cadmium telluride; and high-purity germanium. This figure was created some years ago; the sensitivity of detector materials has improved since then.

Figure 4. Gamma-Ray Spectra of Plutonium-239

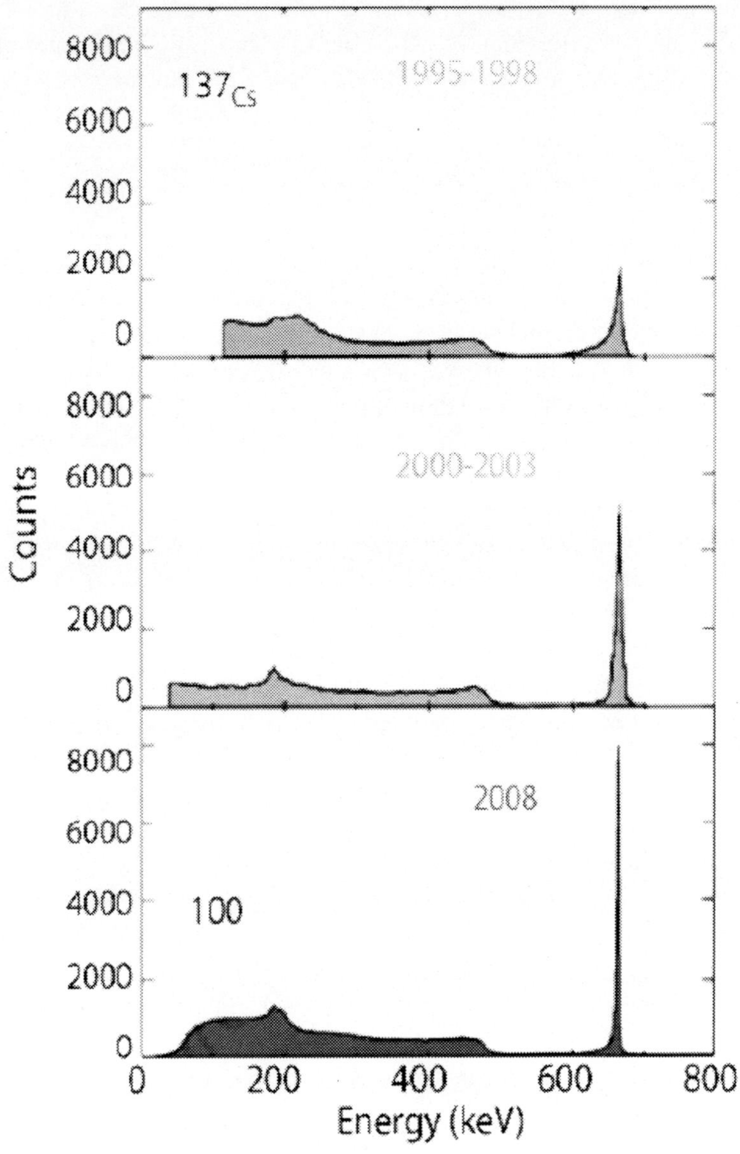

Source: Brookhaven National Laboratory.

Notes: Spectra were taken with cadmium-zinc-telluride (CZT)-based detectors that were high-end in each period listed. The peak on the far right of each spectrum is of particular value for identification of radioisotopes; it becomes sharpers (narrower and higher) with each successive detector. As discussed in the Appendix, many factors were responsible for this improvement.

Figure 5. Resolution of the Cesium-137 Gamma-Ray Spectrum by Different CZT Detectors Has Improved Over Time

Means of detection

Nuclear detection uses neutrons and photons in various ways. Because either neutrons or photons can readily penetrate most materials, they are the main forms of radiation used to detect radioactive material passively, such as by sensing radiation coming out of a cargo container. Gamma rays and x-rays can be used in an active mode to probe a container for dense material through radiography, which creates an x-ray-type image. Neutrons of any energy level, and photons above about 5.6 MeV, can be beamed into a container to induce fission in SNM. Fission results in the emission of neutrons and gamma rays, which can be detected.

Detecting gamma rays

Gamma rays do not have an electrical charge, but an electrical signal is needed to measure them. There are two main ways to turn a gamma ray into electrical energy. One is with a scintillator material, such as PVT. When a gamma ray interacts with this material, it emits many lower- energy photons of visible light. A photomultiplier tube converts them to electrons, then multiplies the electrons to generate a measurable pulse of electricity whose voltage is proportional to the number of lower-energy photons, which in turn is proportional to the energy deposited by the gamma ray. An electronic device called a multi-channel analyzer sorts the pulse into a "bin" depending on its energy and increases the number of counts in that bin by one. A software package draws a histogram with energy on the horizontal axis and counts on the vertical axis. The histogram is the gamma-ray spectrum for that isotope. In contrast, a semiconductor material, such as HPGe, turns gamma rays directly into an electrical signal proportional to the gamma-ray energy deposited. A voltage is applied across the material, with one side of the material the positive electrode and the other the negative electrode. When a gamma ray interacts with the material, it knocks electrons loose from the semiconductor's crystal lattice. The voltage sweeps them to the positive electrode. Their motion produces an electric current whose voltage is proportional to the energy of each gamma ray. Each pulse of current is then sorted into a bin depending on its voltage and the spectrum is computed as described above.

Detecting neutrons

A common neutron detector is a tube of helium-3 gas linked to a power supply, with positively- and negatively-charged plates or wires in the tube. In its rest state, current cannot pass through the helium because it acts as an insulator. When a low-energy neutron passes through the tube, a helium-3 atom absorbs it, producing energetic charged particles that lose their energy by knocking electrons off other helium-3 atoms. Positively-charged particles move to the negative plate; electrons move to the positive plate. Since electric current is the movement of charged particles, these particles generate a tiny electric current that is counted. Neutron count is an important way to identify SNM because SNM and U-238 emit neutrons spontaneously in significant numbers. Few other sources do. Neutron spectra are of little value for identifying isotopes. They do not have lines representing discrete energies, and neutrons lose energy as they collide with low-Z material, blurring their spectra.

Detecting dense material

Photons of high enough energy can penetrate solid objects, but are scattered or absorbed by dense objects (or a sufficient thickness of less-dense material). This is the basis for radiography. For cargo scanning, x-rays or gamma rays are beamed through a container, and a detector on the other side records the number of photons received in each pixel. An algorithm then creates an opacity map of the contents. While a bomb would present a sizable image, dense objects in a container might mask a piece of SNM.

Evasion of detection technologies

An enemy could use various means in an effort to defeat detection systems. One such means is shielding. Gamma rays may travel many feet through such low-Z material as wood, food, and plastic, but high-Z material absorbs and deflects them. Conversely, low-Z material absorbs and scatters neutrons, but they pass more readily through higher-Z material. Different amounts of material are needed to attenuate gamma rays depending on their quantity and energy level. Gamma rays from WGPu are sufficiently energetic and plentiful that they are difficult to shield, while a thin layer of lead would render gamma rays from HEU essentially undetectable,[11]

Sources of radiation other than SNM complicate detection. Background radiation from naturally-occurring radioactive material, such as thorium and uranium, is present everywhere, often in trace amounts. Cosmic rays generate low levels of neutrons. Some commercial goods contain radioactive material, such as ceramics (which may contain uranium) and kitty litter (which may contain thorium and uranium). Other radioactive isotopes are widely used in medicine and industry.

Enemy attempts to defeat one type of detection system may complicate a plot or make it more detectable. (1) It is harder to defeat systems that detect multiple phenomena than a system detecting one phenomenon only. For example, shielding a bomb with lead to attenuate gamma rays would create a large, opaque image on a radiograph. For this reason, Congress mandated that U.S.-bound containers loaded in foreign ports be "scanned by nonintrusive imaging equipment and radiation detection equipment at a foreign port before it was loaded on a vessel."[12] (2) An enemy could attempt salvage fuzing, which would detonate a weapon that sensed attempts to detect it, such as with photon beams, or to open it. However, salvage fuzing could detonate a weapon by accident; if it were scanned overseas; or in a U.S. port, where it would do much less damage than in a city center. (3) Attempts to smuggle HEU into the United States to avoid detection of a complete bomb would require fabricating the weapon inside this nation, which could require such activities as smuggling in other weapon components and purchasing specialized equipment, and could run the risk of accidents, any of which could provide clues to law enforcement personnel.

Current Detection Technologies

The United States deploys various technologies, such as the following, to detect nuclear weapons or SNM. They are available but have important drawbacks.

Radiation "pagers"

These devices, about the size and shape of a pager, can detect radiation at close distance to alert individuals to the presence of elevated levels of radiation. They may use any of several types of detector material. They are lightweight and inexpensive, but cannot identify the material causing an alarm.

Radiation portal monitors

Many of these devices use large sheets of plastic scintillator material, such as PVT, to detect radiation coming from a vehicle. RPMs were deployed soon after 9/11 because they were available at moderate cost. However, PVT cannot identify the source of radiation. Yet many items in everyday commerce contain radioactive material. As a result, some RPMs produce many false alarms, which may require considerable effort to resolve, delaying the flow of commerce. Newer versions have some isotope identification capability.

Radioactive isotope identification devices

These devices are typically hand-held. They have software that can identify a radioisotope by its gamma-ray spectrum. The most capable of these devices use a crystal of high-purity germanium, a semiconductor material, and are considered the "gold standard" of all identification devices. Such devices are heavy and delicate, and must be cooled with liquid nitrogen or by mechanical means, limiting their usability in the field. They have a relatively short range for detecting radiation sources with low radioactivity, notably shielded HEU, making them unsuitable as the primary method of screening cargo containers.

Radiographic imaging systems

These devices send high-energy photons through cargo containers to create a radiographic image of the contents. The radiograph is scanned, either automatically or by an operator, to search for nuclear weapons, contraband, stowaways, and other illicit cargo. While a nuclear weapon would show up as a white (or black) image on the radiograph and would be clearly visible if hidden in a shipment of low-Z material like food or paper, an operator might overlook it if it were in a shipment of other large, dense objects or jumbled items of various sizes and densities. A small piece of SNM might also be overlooked.

2. Advanced Technologies

Many nuclear detection technology projects are under way in the United States and elsewhere. This section discusses nine U.S. technologies selected to include different (1) *agencies sponsoring projects* (the Defense Threat Reduction Agency (DTRA), an agency of the Department of Defense (DOD); the Domestic Nuclear Detection Office (DNDO), an agency of the Department of Homeland Security (DHS); and the National Nuclear Security Administration (NNSA), an agency of the Department of Energy (DOE)), (2) *organizations performing the work* (national laboratories, industry, universities, and collaborations between them), (3) *types of technology* (materials, algorithms, simulation, systems), (4) *physical principles* (muon tomography, radiography, stimulated emission of radiation, nuclear resonance fluorescence), (5) *distances* between the detector and the object being scanned (near and far), and (6) *levels of maturity* (in use for many years but constantly updated, near deployment, and anticipated to be available for deployment in several years). This section does not consider technologies in early development because it is too soon to tell how they will pan out. The discussion of each technology includes several categories:

- The problem that the technology addresses
- Technology background
- Description of the technology
- Potential benefits that the technology offers
- Status, schedule, and funding
- Scientific, engineering, cost and schedule, and operational risks
- Potential gains by increased funding
- Potential synergisms

The last three categories require some explanation.

Risks: The discussion presents potential benefits of the technologies. It does not present "cons." That would be premature because development programs address problems. Instead, each section discusses risk. There are several categories of risk. Scientific problems may thwart a technology. Even if it is scientifically sound, it may be hard to engineer into a workable system. Even if it can be engineered, it may be unaffordable, or take too long to develop. Even if it can surmount these hurdles, problems encountered in field use may render it unattractive.

Potential gains by increased funding: In preliminary discussions, project managers asserted they would, if given more funds, use the added funds to solve problems or exploit opportunities. CRS therefore asked managers of all nine technologies considered in this report how they would spend more money as a way to probe for problems and opportunities with their projects. Would they pursue several promising routes to a technology instead of pursuing only one at the outset? Would they buy more equipment so they could speed up the work? Would they hire more staff?

This analysis applies only to the nine technologies discussed in this report. It is not intended to focus authorization or appropriation consideration solely on them. Other technologies not considered here might realize larger (or smaller or no) gains through increased funding.

Potential synergisms: As the technology descriptions show, many systems have common elements, such as certain types of detectors or algorithms, and work on one technology may contribute to others or have applications beyond current plans.

This report now discusses each of the nine technologies using the above format.

Nanocomposite Scintillators[13]

The problem

Scintillator materials are used to detect, and in some cases identify, gamma rays. Higher-performance scintillators are more expensive, harder to manufacture, and fragile; scintillators that are less costly, easier to manufacture, and more rugged offer lower performance. At issue: can the desirable qualities of each type be combined to achieve better performance at lower cost?

Background

Scintillators are materials that, when struck by photons of higher energy, such as gamma rays, capture this energy and release it as photons of lower energy, usually visible light. The material should capture as much of the energy of each photon striking it as possible in order to build an accurate photon energy spectrum and thus identify the material emitting the photons. Ideally, a photon should deposit its full energy in the scintillator material, a so-called full energy interaction. It is also important that the deposited energy be efficiently converted into photons of visible light, which are then counted to determine energy.

There are two main types of scintillators. Inorganic scintillators (those not containing carbon) are typically single crystals, such as sodium iodide (NaI) with a small amount of thallium added. The probability of full energy interaction increases sharply with atomic number (Z) of the scintillator material,[14] and is high for inorganic crystals. The more energy from each photon a scintillator absorbs and then gives off, the better the correlation between energy input and output, and the more precise the spectrum that can be constructed. As a result, a device using an inorganic crystal has a good ability to identify the radioactive material producing a gamma-ray spectrum. There are several drawbacks. The area of a detector that is sensitive to gamma rays is small (limited to the size of a crystal), so the detector must be close to the object to be searched or must scan for longer time so it can receive more gamma rays. They are fragile; dropping one can destroy it. Many inorganic crystals absorb water and are sensitive to light, so they must be protected from environmental conditions. NaI crystals are easy to grow, but cost about $3 per cubic centimeter (cc) of crystal. Other, higher-resolution scintillators are harder to grow, are more costly, and are sensitive to moisture. For NaI, light output varies strongly with temperature, so the temperature must be stabilized or the data corrected.

Organic scintillators have the opposite set of properties. They can be made of plastic, such as PVT. As such, they are easy and cheap to make, and are much less fragile than crystals. They can be produced in bulk, making them suitable for deployment in large sheets, such as for radiation portal monitors. On the other hand, since they are composed mostly of hydrogen and carbon, both very low Z elements, they are very inefficient at absorbing the total energy of gamma rays. As a result, as **Figure 3** shows, peaks in PVT-produced gamma ray spectra are indistinct at best, making such spectra of little or no value for identifying radioisotopes.

Technology description

A research project under way at Los Alamos National Laboratory, the Nanocomposite Scintillator Project, seeks to combine the advantages of both types of scintillator materials to overcome the disadvantages of each. The principle is that "nanocrystals," crystals 2 to 5 nanometers in diameter (1 nanometer = 1 billionth of a meter), of certain inorganic scintillator materials can capture most of the energy from photons, thus offering nearly the performance of single large crystals, if packed densely enough in plastic. The resulting mixture also has the desirable features of plastic. The crystals are lanthanum bromide mixed with cerium, or cerium bromide. The modified polystyrene plastic is a scintillator material, so it increases the amount of energy converted to a detectable signal.

In operation, when a gamma-ray photon strikes this material, its energy is absorbed by nanocrystals and the plastic, raising some atoms to a higher energy level. These atoms give off this energy as photons in the visible and near-visible regions of the electromagnetic spectrum ("optical photons"). A photodetector, an electronic component that generates many electrons for each photon it receives, amplifies the signal and converts it from an optical signal to an electronic signal that can be counted. The number of optical photons generated corresponds to the energy level of the photon striking the material. A multi-channel analyzer counts the optical photons, determines the energy level of the photon striking the material, and increases the count of photons of that energy level by one, by this process creating a gamma-ray spectrum.

Edward McKigney, the principal investigator, believes that nanocomposite scintillator material will be able to discriminate between neutrons and gamma rays. He asserts that simulations support this case. The project has obtained data from experiments using the plastic without nanocrystals and nanocrystals without the plastic, and from the literature, and has used these data in simulations. Basic physics calculations also support this case. Neutrons generate protons when they strike the plastic, and gamma rays generate electrons. The plastic responds differently to protons and to electrons; the same is true of the nanocrystals. However, as of August 2008 the project had not conducted experiments demonstrating the ability of the material to detect neutrons and to differentiate between them and gamma rays.

Potential Advantages

McKigney states that because the crystals are tiny, growing them is not difficult and is much less costly than growing large whole crystals of these materials. The material can be fabricated at an industrial scale, he says, further reducing cost. He estimates that this material potentially offers the performance of $300/cc material (e.g., lanthanum bromide) at a cost of 50 cents/cc. Because the material acts as a plastic, it is rugged and flexible, and can be made in large sheets, according to McKigney, increasing the sensitivity of a detector using it. It would operate at ambient temperatures.

Status, schedule, and funding

The project has several components. The largest is to develop nanocomposite scintillator material. There are several smaller components: process scale-up; electronics development; detector design; and basic research for a more advanced material. The budget for the entire project is as follows. Early research started in 2004. Los Alamos provided $65,000 in initial funds in FY2005. The laboratory and DNDO provided $1.2 million for FY2006 and $1.6 million for FY2007. Los Alamos, DNDO, and DTRA provided $4.6 million for FY2008 and are projected to provide $5.5 million for FY2009. The project seeks to deliver a small cylinder (1 inch in diameter and 1 inch high) of the material, and to characterize its performance, by December 2008. Anticipated costs are $4 million for FY2009 for the nanocomposite scintillator component. The goal for the end of FY2009 is to have a pilot-scale demonstration of the scintillator material and to start transferring its technology to industry. This material would be optimized for gamma-ray detection. A constant level of effort of $4 million per year after FY2009 for two or three years would be used to optimize the material for combined neutron and gamma-ray detection. As of August 2008, NNSA's Office of Nonproliferation Research and Engineering (NA-22) is considering a proposal from the Nanocomposite Scintillator Project to start the combined gamma-ray and neutron detector material development in FY2009 so it can proceed in parallel with the existing program.

Risks and concerns

Scientific risks and concerns

Fabrication of this material requires finding chemicals that can coat the surface of the nanocrystals so they can disperse properly in the plastic while optimizing other properties.[15] The nanocrystals must be packed densely in the plastic to increase sensitivity and resolution. As of May 2008, researchers had reached a packing level of 6% (by volume), with a goal of 50%. It remains to be seen if they can meet this goal. Packing crystals densely in plastic will change some properties of the plastic,[16] so care must be taken to minimize this problem. Another source of risk is that, as of August 2008, the project had not reached high enough packing levels to allow for measurements to determine the sharpness of gamma-ray peaks in spectra generated by this material. The project plans to conduct experiments on this point by December 2008. Such measurements are expected to reduce scientific risk. However, unexpected nanoscale physics could impair energy resolution. In that case, the particle structure would have to be engineered to mitigate the effects; at worst, these effects might degrade performance.

Engineering risks and concerns

(1) When developing the plastic-crystal composite, attention must be paid to ensuring that the material can be made with industrial processes used to manufacture other plastics. (2) The chemical reaction that occurs in manufacturing plastic gives off heat, potentially creating hot spots that would impair the performance of the material. This is not a problem for very small quantities. At the other end of the scale, for industrial production, the problem is well understood and chemical engineering solutions have been implemented for decades. A concern is whether a solution can be devised for pilot-scale production.

Cost and schedule risks and concerns

The project is not mature enough to provide a firm estimate of the cost of the material produced on an industrial scale. The cost estimate cited above is based on the cost of procuring the raw materials and assumes that the energy cost for processing is low. Inflation in energy would not be expected to increase cost sharply, but cost of the product is sensitive to the cost of cerium, a rare earth element. The manufacturing processes are similar to those used for such consumer goods as fabric softeners and disposable plastic water bottles, so unit cost arguably should not be high. However, the cost estimate excludes the cost of fixed infrastructure; how much that would add to unit cost depends on infrastructure cost and the number of units produced. Transferring the technology to a commercial partner by the end of FY2009 depends on resolving potential scientific and engineering problems in a timely manner and finding the right partner. The project is not far enough along to make a firm estimate of schedule beyond FY2009.

Operational risks and concerns

In any project, it is possible to develop a product only to have it fail in everyday use. This project is trying to minimize this risk. It is trying to design robustness into the material, such as by (1) using a plastic that is soft and rubbery rather than brittle, (2) ensuring that the material will be effective across the temperature range to which it will be exposed, and (3) ensuring that the detector and packaging are compatible so that, for example, the detector material will not expand so much as to crack its casing.

Potential gains by increased funding

The project's total budget is about $5.5 million per year. McKigney states that he could "usefully employ a total budget of up to $12M/year to reduce the time to deployment of these technologies" in several ways. (1) The project is pursuing, with about one-third of its total funding, a separate basic research project to develop a scintillator material approaching the resolution of high-purity germanium detectors with the cost and processing characteristics of plastic. More funds could accelerate this project, according to McKigney. He cautions that this project might take a decade or more and has much greater scientific risk than the current nanocomposite scintillator project. (2) The project uses equipment available at Los Alamos, but he states that staff could save time if they had their own equipment.[17] (3) It would be difficult to fabricate a single large panel (e.g., 50 cm wide by 20 cm thick by 90 cm long) of detector material. Further, a panel segmented into tens to thousands of tiny panels produces the optimum tradeoff between cost and effectiveness and can provide data on the position of a radioactive source. Each panel element would need its own electronics channel, so application-specific integrated circuits must be custom-designed, which could take two or three years. With more funding, he asserts, the project could develop these chips and the scintillator material concurrently, saving time. Recognizing this leverage, DTRA provided about $150,000 for this purpose spread over 2-1/2 years beginning in May 2008. McKigney states that additional funding would accelerate this schedule. (4) Hiring more chemists, electrical engineers, and others would accelerate these projects, according to McKigney.

Potential synergisms and related applications

(1) This material offers the greatest benefit in detectors that use large panels of scintillator material, such as some under development as discussed below. (2) By offering greater sensitivity and greater resolution, this material could provide better data for algorithms to process, permitting the development of simpler, more capable algorithms. (3) Nanocomposite scintillator material may be able to measure neutron and gamma ray energies. As such, it might be possible to use it in a detection system instead of separate means of detecting each particle type. (4) Some companies are considering systems that use tubes filled with helium-3 gas to detect neutrons. However, helium-3 is scarce, and there may not be enough to support large-scale use of these tubes. Nanocomposite scintillator material might be an alternative for neutron detection.

GADRAS: A Gamma-Ray Spectrum Analysis Algorithm[18]

The problem

In a carefully-controlled laboratory environment, a radioisotope can be readily identified by matching its gamma ray spectrum to one in a library of spectra. At a port or border crossing, a spectrum would be complicated by radiation from many sources. At issue: How can the signal from SNM be extracted from a gamma ray spectrum, and how can this capability be improved?

Background

Nuclear detection hardware receives much attention, but data from the hardware—e.g., pulses of various energies from gamma rays—are unintelligible until processed by software. Indeed, software in the form of algorithms is the "brains" of a detector. An algorithm, such as a computer program, is a finite set of logical steps for solving a problem. For nuclear detection, an algorithm must process data into usable information fast enough to be of use to an operator.

Algorithms are designed to assure a low rate of false positives, which impede commerce, and a near-zero rate of false negatives, which could let a terrorist weapon into the United States. For gamma-ray identification, an algorithm creates equations to model "radiation transport," the movement of radiation through material. A detector will record gamma rays from all sources— background radiation, items in ordinary commerce, radioisotopes that terrorists might include in a shipment to confuse analysis, and a nuclear weapon or SNM. Many uncertainties affect an algorithm. Gamma rays lose energy as they interact with matter, altering their spectra depending on the type and thickness of the materials they pass through, and the initial energy of a gamma ray. Random fluctuations in the number of counts in each gamma-ray energy bin create statistical uncertainties, especially if the number of counts is low. The composition and thickness of materials between the radiation source and the detector (cargo, shielding, container wall, air, etc.) is not known. The equations may not exactly represent the detector dimensions or the shielding configuration because approximations are made to reduce complex environments to a set of equations that can be more readily computed.

Technology description

Dean Mitchell of Sandia National Laboratories has developed an algorithm, "Gamma Detector Response and Analysis Software," or GADRAS.[19] Development of GADRAS started in 1985 for use in the Remote Atmospheric Monitoring Project (RAMP), which used low-resolution detectors to analyze airborne radionuclides.[20] Beginning in 1996, automated isotope identification was developed within GADRAS to process spectroscopic data collected at border crossings to support interdiction of radioactive materials.[21] Work on the current version of GADRAS started in 2001. In earlier versions, it was seen as acceptable to spend a month analyzing an individual spectrum; in the wake of 9/11, in order to be of value for screening commerce, GADRAS had to be modified to process data quickly while minimizing false positives and false negatives.

To identify radioactive sources creating a gamma-ray spectrum, GADRAS matches an entire gamma ray spectrum against one or more known spectra. Many other algorithms focus on peaks in gamma ray spectra because they are the most obvious features. In contrast, GADRAS analyzes the full spectrum for several reasons. (1) Peaks may overlap, making source identification ambiguous. (2) Most counts in a gamma-ray spectrum are often outside the peaks, in which case using only peak data would ignore most of the data. For example, less than 3% of the counts in a spectrum for U-23 8 occur in the 1.001 -MeV peak, the most prominent feature of its spectrum. (3) Counts outside the peaks help characterize the composition and thickness of intervening material. Since gamma rays interact with these materials, characterizing the materials improves the accuracy with which the gamma-ray spectrum as read by a detector can be linked back to the gamma-ray source. Arriving at a solution consistent with all the data increases confidence in the result. (4) The absence of counts in a region of a spectrum can be a clue to the identity of radioactive materials. (5) Using the entire spectrum helps analyze data from scintillators having low energy resolution because low resolution often precludes identification of peaks in the spectrum, and helps analyze spectra of weak sources.

GADRAS also uses neutron flux data. Since neutrons pass more readily through high-Z material and gamma rays pass more readily through low-Z material, different materials, such as in a container, will affect the total radiation output differently. As a result, using both gamma ray and neutron data improves the analysis.

GADRAS has been in use since 1986. Since 9/11, more operators have used it in a wider range of applications. In response, the software is continually upgraded to support new types of radiation sensors, provide new capabilities, and meet new performance requirements. One goal is to make it more automated so that less skill is required to operate it. Another goal is to make it faster. A current effort focuses on improving sensitivity to SNM and reducing the false alarm rate. In the past, some sacrifices were made to fidelity of the analysis in order to gain speed; now, with faster computers, it should be possible to improve the analysis while increasing speed; this approach is being investigated. Another approach to reducing false alarms is to examine radiation sensor data collected in 2005 on cars entering the Lincoln Tunnel between New York City and New Jersey. According to Mitchell, the data had 50 to 60 false alarms. Mitchell and other researchers at Sandia are trying to determine what caused the false alarms in order to modify GADRAS to correct for these problems.

While parts of GADRAS are used to a limited extent in the field, its main use is for "reachback" analysis. A radiation detection operator in the field, such as a Customs and Border Protection (CBP) officer, who finds a vehicle or cargo container that presents a suspicious radiation signature that cannot be easily resolved, can send the detection data (such as a gamma ray spectrum) to the Laboratories and Scientific Services section of CBP for a more detailed analysis.[22] That analysis uses GADRAS. Similarly, if that service is unable to resolve the matter, it can send the data to a secondary reachback at the nuclear weapons laboratories, which also use GADRAS.

Potential advantages

By analyzing a complete gamma ray spectrum, GADRAS increases the accuracy of determining whether a cargo container or other item is carrying SNM, reducing the risk of false positives and false negatives. Using neutron count data in addition to gamma ray spectra further improves capability.

Status, schedule, and funding

GADRAS has been used for cargo inspection since 1998. Software upgrades are released every two months or so. There is no line item for GADRAS development. Mitchell estimates that Sandia is spending perhaps $500,000 in FY2008 to continue to develop GADRAS. DTRA states, "The DTRA in concert with NNSA is currently proposing development of the next generation of GADRAS as part of a potential [memorandum of understanding] between the organizations. The effort would emphasize the revision algorithms, update to Vista compliant software, and increasing portability for field use."[23]

Risks and concerns

Scientific risks and concerns

Since upgrades are ongoing, developers face such scientific problems as how to improve equations to characterize the response of detector material to radiation. Such problems are not a major risk to continued development of GADRAS, Mitchell states. Another risk is that the gamma-ray signal from shielded HEU may be so low that even a high-quality algorithm cannot identify the HEU.

Engineering risks and concerns

The major risk to GADRAS development comes from the programming language that is used for the graphical user interface (GUI). The GUI for the current version of GADRAS is written in a Microsoft language called Visual Basic Version 6 (VB6). VB6 functions under current versions of Windows, including the most recent, Vista. Microsoft, however, no longer maintains VB6, so it is not necessarily compatible with new compilers,[24] often leaving no effective upgrade path for large application programs like GADRAS. User feedback indicates that some GADRAS components do not function properly in Vista. While minor changes may resolve the latter problem, the current GUI may not run under a future version of Windows.[25] GADRAS is a large program, and current funding does not support the effort that would be required to update it.

Cost and schedule risks and concerns

Since upgrading GADRAS is a low-budget activity, it entails little cost risk. No near-term requirements impose significant schedule risks. However, it would take one to two years to revise the GUI. Making this revision in advance of a change of operating systems would enable users to use GADRAS without interruption when they switch to the new operating system.

Operational risks and concerns

Since GADRAS has been in existence for many years, and since it is used mainly by scientists and technicians rather than by operators in the field, there is little risk that it will fail in everyday use. The risk that an upgrade will cause a problem is reduced by careful testing. Another risk is the possibility that GADRAS would not run on Vista or a future operating system and that conversion would not be made before an operating system that would not run GADRAS is introduced. Some have said that GADRAS is a complicated program to use; accordingly, the CONOPS for GADRAS is that it is used mainly for reachback.

Potential gains by increased funding

Mitchell states that perhaps another $1 million a year for two years would allow Sandia to hire a few programmers who could convert GADRAS to use current compilers in order to avoid potential disruptions associated with new operating systems.

Potential synergisms and related applications

(1) GADRAS might improve the performance of systems that induce fission and detect the resulting radiation. (2) By gaining more information from the gamma ray spectrum, it

might reduce the gamma ray or neutron flux needed to induce fission and the dose resulting from fission, thereby increasing worker safety. (3) More computation power would permit more sophisticated iterations of GADRAS to be developed, or would permit GADRAS to run faster, or both. (4) Increased computation power, especially smaller and more capable computers, might enable GADRAS to be modified for use in the field with radiation detection equipment, permitting quicker resolution of suspicious containers and vehicles. Reachback would then be used only for the hardest-to-resolve cases. (5) GADRAS could be modified to improve the performance of radioisotope identification devices (RIIDs). DTRA states that it "is currently funding development of a more portable version of GADRAS ... that is intended to be resident on certain RIID systems, such as handheld HPGe system produced by Ortec."[26]

Computer Modeling to Evaluate Detection Capability[27]

The problem

Developing equipment to detect terrorist nuclear weapons and SNM requires many choices. It would be of great value to evaluate how they affect cost and performance before committing to a system. However, many combinations and tradeoffs are possible, and it would be prohibitively expensive and time-consuming to run enough trials for each to make a valid comparison among them. When faced with similar choices, such as in designing a car, corporations typically run huge numbers of simulated trials using computer models that take significant investment to develop and maintain. At issue: How can computer modeling contribute to the development of nuclear detectors, and what are its limitations?

Background

A detector should maximize the probability of detecting an actual threat (a true positive) while minimizing the probability of detecting a nonexistent threat (a false alarm, or false positive). For a given technology, these objectives cannot be achieved simultaneously—an improvement in either one comes only at the expense of the other. A receiver operator characteristic (ROC) curve, such as **Figure 6**, illustrates this relationship.[28] By relating the true positive and false positive rates, the curve defines the performance of a receiver (in signal processing, where the term "ROC" originated) or of nuclear detection equipment. ROC curves show that the probability of a true positive and a false positive go up together. This is logical; one could eliminate false alarms by turning off the detector, or could be sure of detecting every threat by having the detector alarm whenever the slightest trace of radiation is detected, which, given omnipresent background radiation, would be all the time. The peril of failing to detect an actual threat is clear. At the same time, law enforcement and commercial interests are intolerant of false alarms because these alarms require a major effort, diverting officers to the scene and possibly unloading a cargo container or closing a border crossing. Further, in the real world, numerous false alarms may cause operators to ignore all alarms or set the detector to be less sensitive, reducing the false alarm rate but also increasing the likelihood of missing an actual threat. This tradeoff is shown in **Figure 6** by moving from point C, with a high probability of detection but a high false alarm rate, to point B, with intermediate values for both, to point A, with a low false alarm rate but a low probability of detection.

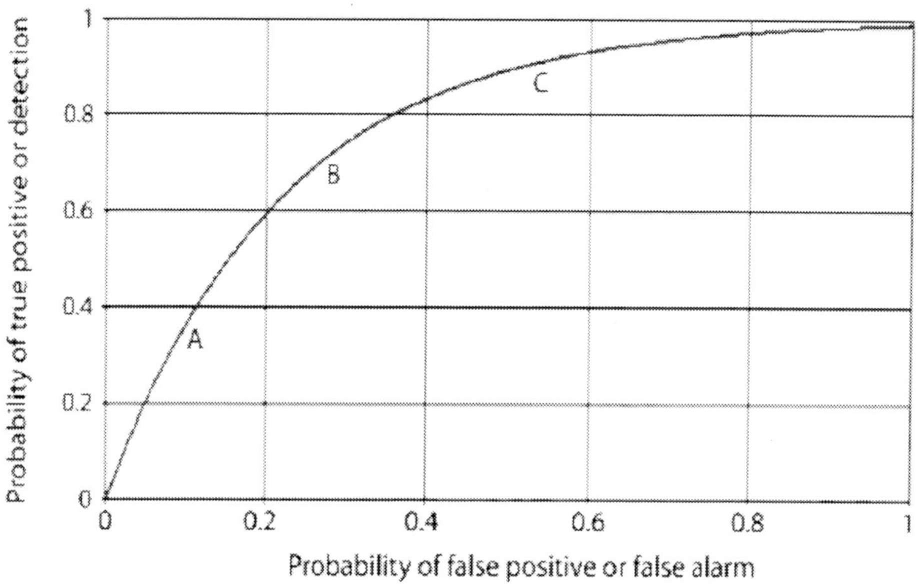

Source: Prepared by CRS with the assistance of Lawrence Livermore National Laboratory.

Figure 6. A Notional Receiver Operating Characteristic (ROC) Curve

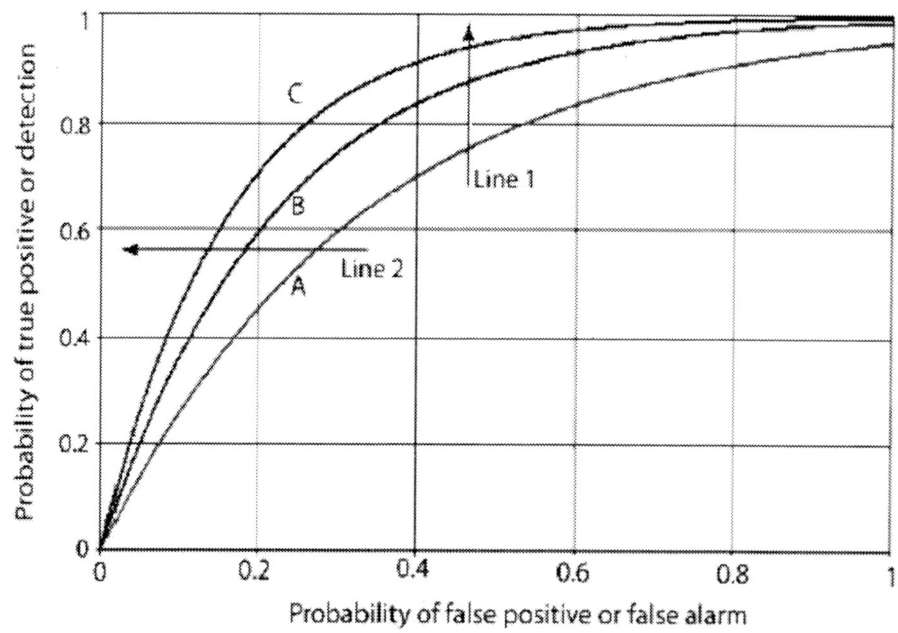

Source: Prepared by CRS with the assistance of Lawrence Livermore National Laboratory.

Figure 7. Three Notional ROC Curves

Figure 7 shows three ROC curves to illustrate several concepts. (1) Moving from curve A to curve B to curve C shows the performance of a hypothetical detector improving over time, perhaps as a different detector material is used or an algorithm is modified. The improvement can be visualized by moving upward (line 1), which shows an increase in the true positive rate for a given false positive rate, or by moving from right to left (line 2), which shows a reduction in the false positive rate for a given true positive rate. (A diagonal line from lower right to upper left would show improvement in both.) (2) A, B, and C could represent differences in performance of one detector under different conditions, such as changes in the background, different operating conditions (e.g., scan time), or different benign materials in a container. (3) The curves could characterize the performance of three competing detectors. Note that actual ROC curves have more complex shapes than the notional curves shown. They may even cross over each other, indicating that one option is not uniformly better than another, requiring consideration of further tradeoffs.

Many variables affect detector performance. Some are operational, such as scan time, the detector's target (containers, cars, trucks, trains), the distance between detector and target, and environmental conditions (background radiation, season, temperature). The detector is to detect SNM or nuclear weapons, yet the signatures on which it will focus may be accompanied by radiation from innocent sources and background radiation, and may be shielded inadvertently or deliberately. There are choices for the active elements of a detector, the algorithm used, and specific subroutines. These choices affect detector performance. Many combinations of these variables are possible. To gain enough data to make a ROC curve statistically valid, many trials would need to be performed for each combination of controllable variables (operating setup, detector, and algorithm) against many targets, each generating its own radiation signature. Each event in which a vehicle or container passes through the detection system is called an "encounter." It would be prohibitively expensive and time-consuming to run many encounters for each of thousands of combination of controllable variables, but it would be of great value to have the resulting data in order to compare, improve, or choose between detection systems and their components.

Computer modeling can help. Modeling creates mathematical representations of the real world, varies the inputs, and calculates the outputs. In the case of nuclear detection, the real world consists of controllable variables (operational scenario, detector, and algorithm) and uncontrolled variables (signals from radioactive material). The performance of a modeled detector can be illustrated using ROC curves as described above. Running the model to simulate many encounters for each combination of controllable variables provides enough observation points to generate a statistically significant ROC curve. This process is repeated for many combinations of variables. The resulting data show the user how change to variables affects performance, and permit comparing one detector against another. Computer-generated "data" for nuclear detection encounters are always imperfect, as discussed below. As a result, the adequacy with which the models represent reality is always at issue, and model developers devote great effort to improving that representation.

Technology description

DNDO has established an ongoing program, Detection Modeling and Operational Analysis (DMOA), that the national laboratories and private sector contractors carry out. It builds models that characterize the variables noted above, i.e., the operational scenario, the radiation signals, the detector, and the signal-processing algorithm. Creating mathematical representations of the first three takes an immense amount of work because each is so complex. Alternative algorithms are simulated as part of the overall detection performance modeling. (Algorithm development requires a great deal of work, but is not within the scope of DMOA, which focuses on simulations.) For example, modeling a gamma-ray spectrum requires taking into account various sources of radiation and types of shielding. A DMOA study of real-world data found that the spectra from cargo differed between spring-summer and fall-winter; the study speculated that a different mix of products shipped in the two periods caused the difference.[29]

The model processes data on the operational scenario, radiation signals, and the detector to create a gamma-ray spectrum that is sent to the algorithm. The algorithm does not "know" the difference between a spectrum generated in the real world or by computer, and processes both in the same way. Since the difference is the source of the data, the key to simulation is generating the gamma- ray spectrum (signal plus noise) that goes to the algorithm. The greater the fidelity with which the model mimics real-world inputs, the better it represents system performance. Modeled performance, as measured by detection probability and false alarm rate, can be summarized in a ROC curve in the same way as empirically-measured performance. The modeling effort also includes assessing the sensitivity of detector performance to changes in operational scenarios, detector hardware, and algorithms.

One DMOA effort in 2006 focused on four main areas to improve its models.[30] This effort illustrates the work undertaken to improve the fidelity of models and how this work requires detailed knowledge of the components being modeled.

- Evaluation benchmarks. Providing a basis for comparing systems requires a reference set of threat objects and shielding. Previously, this set included plutonium, HEU, and other threats, as well as different levels of shielding. In 2006, DMOA added new threats and shieldings to the reference set. It also developed a reference set of objects for detection by radiography.
- Background and nuisance source modeling. According to DMOA, "Background radiation and nuisance source population characteristics generally dictate detection threshold settings through their impact on innocent alarm rates. Characterization of these factors is critical to evaluating the performance of radiation detection systems." DMOA used real-world data to develop a model of the distribution of naturally-occurring radioactive material. Data from the model were then compared with another set of real-world data to check the validity of the model.
- Operational analysis. DMOA used real-world data to compare the performance of several algorithms. One result was to "highlight the sensitivity of system performance to the detection algorithm used."
- Detector resolution study. DMOA studied the resolution needed to distinguish threats from other sources of radiation. The analysis, though preliminary, found that there is a benefit by having better resolution than that offered by sodium iodide, but that improving resolution beyond a certain point would offer marginal benefit.

Potential advantages

Modeling offers advantages compared to obtaining data from the real world. (1) A model permits construction of statistically significant ROC curves. It is not uncommon to require a false alarm rate of one in ten thousand or lower; statistical validation of performance at this level would require hundreds of thousands or millions of experiments. It would be costly and time-consuming to conduct field trials using a single combination of variables in order to build one statistically significant ROC curve, let alone conducting field trials for thousands of combinations of variables that represent the range of conditions. A model permits running many thousands of simulated encounters in less than a day to explore the range of encounters a detector may face. Once the model has been constructed, validated, and implemented, the cost of these computer runs is very low. (2) A model and a test program are complementary. The test program generates data for the model, and the model can steer the test program by indicating what tests would be of greatest value for improving the predictive power of the model. (3) A model permits exploration of encounters that might occur in the real world but that could not be conducted because of cost, safety, or difficulty. For example, it would be unthinkable to place a nuclear weapon in commercial traffic to test detectors, but the only facilities where such testing could be done, notably the Nevada Test Site, have characteristics very different than those of ports or border crossings. As another example, it might be desirable to see whether a protracted period of heat or high humidity would impair how a detector would function, but it would be much easier to replicate such conditions through modeling. (4) A model permits comparison of components of an encounter, such as which algorithm better processes the data for a given gamma-ray spectrum. (5) A model permits analysis of alternatives and clarification of tradeoffs before committing to a specific design, helping to inform billion-dollar decisions.

Status, schedule, and funding

DMOA is an ongoing program. When DNDO was established, DMOA became an explicit element within the System Architecture program. It has been funded within that program at approximately $2 million per year. Other detection modeling is done for other purposes by other offices within DNDO, as well as by DOE and DOD. However, since modeling activities are inherently cross-cutting and support many technology development and assessment projects, it is difficult to estimate total spending on modeling in the federal budget.

Risks and concerns

Scientific risks and concerns

While no model can be perfect, the key risk is that the model's output (for example, gamma ray spectra that the model generates) might differ significantly from data that would have been obtained from actual field trials. Part of this risk concerns modeling the underlying science. Some aspects are known in detail, such as the radiation spectrum of U-235. But there are uncertainties. HEU is not pure U-235, so the spectrum will be somewhat different from that of U-235. Some shielding can be modeled precisely, such as a centimeter of lead. But the defense (the modeling design team) does not know what shielding, if any, the attacker might use. A cargo container can hold many types of cargo, each of which interferes with radiation transport differently. Different arrangements of the cargo place different amounts and types of material between source and detector, affecting gamma ray spectra. Any model makes approximations and simplifications, sometimes to allow the simulations to run faster, or simply because more fidelity for certain phenomena are judged unnecessary or unwarranted. This creates further risk that the model might not sufficiently represent reality. There are other risks. Items included in the model may not be selected accurately. Systematic errors may arise, such as differences between spring-summer and fall-winter cargo. Adding detailed radiation sources with additional shielding, types of cargo, and detector details may increase the realism of the model, but also add complexity, opening the door to additional errors.

Engineering risks and concerns

Models are complex mathematical approximations of reality. Yet real-world data are often limited, perhaps covering too narrow a range of conditions. DMOA uses statistical methods to transform a real-world data set into data for different detectors and conditions. As a DMOA report states, "LLNL [Lawrence Livermore National Laboratory] has developed a procedure for generating a statistical model of a nuisance source population ... based on measured data. ... The statistical model developed provides a basis for simulating an unlimited number of random samples in nuisance sources for a population assumed to be similar to that underlying the measured data."[31] A difficulty is in validating the model. Various isotopes cause background radiation, and DMOA observes that "there is no simple standard procedure to compare multivariate populations."[32]

Models may also manipulate data by adding in, or "injecting," spectra from well-characterized radiation sources like HEU or WGPu to computed spectra of cargo typical of normal commerce to see how well an algorithm can detect threat material in cargo. However, that approach may be an inadequate representation of reality. It is arguably unlikely that terrorists would include a nuclear weapon or SNM in a random cargo container if they chose that vector; rather, they might arrange the cargo to help evade detection. Simulation developers have not modeled such deliberate arrangements of cargo because that would require thinking about how to model adversary behavior, something outside their expertise. It might be desirable to have terrorism experts modify some injections to reflect adversary behavior.

Cost risks and concerns

Most funds spent on DMOA are for staff salaries, so the likelihood of unanticipated cost increases appears low. However, this assumes that computing hardware required to run the models quickly is already paid for.

Schedule risks and concerns

If terrorists planned to bring a nuclear weapon or SNM into the United States, they would presumably try to evade detection, leading to an offense-defense competition. It would be essential for this nation to stay ahead. That task involves much effort by many entities. For modeling, the task would be to upgrade means of evasion included in the model in time to stay ahead of the threat. This would depend in part on inputs from intelligence services, such as data on specific threats, but also on advances in characterizing background radiation, developing algorithms, developing statistical means of transforming or validating data, and the like. The program is paced by resources available. The risk is that progress in nuclear detection, including modeling, will not be fast enough to defeat the threat.

Operational risks and concerns

Law enforcement and commercial interests are intolerant of false alarms. The false alarm rate often drives detector settings and choice of detection systems, and increasing the true positive rate also increases the false positive (false alarm) rate. While a model can explore the tradeoff and optimum balance between true and false positive rates, a risk is that systematic errors, incorrect data transformations, and incomplete accounting for background radiation would cause the model to generate a setting that is less than optimal. In some cases, this problem can be overcome by making adjustments in the field, but in other cases, such as the choice of detector material, adjustments may not be possible.

Potential gains by increased funding

At present, the DMOA efforts at the national laboratories and private sector contractors are distributed throughout their organizations. Most staff members who work on detection modeling do so only part time. Some modeling tools (simulation codes, databases, and algorithms) tend to be somewhat informal, developed by various groups for specific needs. According to Richard Wheeler, Lead for Homeland Security Analysis, Global Security Directorate, Lawrence Livermore National Laboratory, added funding in this area could enable the designation of full-time staff, development of standardized modeling tools, acquisition of dedicated computing resources, establishment of rigorous peer review of the models, formal coordination of related efforts sponsored by DOE, DOD, and other agencies, and generation of new databases essential for model validation. While analysis of detector performance against realistic threats is sensitive or classified, many advances in modeling and modeling tools could be made in an open environment, including the academic community. Wheeler states that more funds could support a more substantial engagement of university researchers.

Potential synergisms and related applications

Simulation does not detect anything by itself, but is by its nature synergistic with other aspects of nuclear detection. It has as its purposes improving many elements of detection systems, whether deployed or under development; helping to integrate hardware and software into a system; optimizing systems and CONOPS; and informing decisions on the most cost-effective mix of systems to acquire.

L-3 CAARS: A Low-Risk Dual-Energy Radiography System[33]

The problem

Current radiography systems have important limitations in their ability to detect SNM in cargo containers. A small amount of dense material may be inconspicuous if shipped in a container filled with dense or random objects. SNM could also be placed inside dense objects for camouflage. Another difficulty of detecting SNM in containers is that an operator may fail to see a threat. At issue: Can radiography take advantage of different characteristics of SNM to detect it in cargo? And can a system evaluate radiographic images automatically to reduce dependence on operator judgment?

In an effort to overcome these limitations, DNDO is conducting R&D on advanced radiography systems under the Cargo Advanced Automated Radiography System (CAARS) program. CAARS involves three technical approaches to radiography, each with a different contractor. This section and the next two discuss these systems. It must be emphasized at the outset that these systems are under development. As a result, no actual CAARS systems exist, so diagrams and performance specifications presented in these sections must be viewed as goals that are yet to be demonstrated in deployable equipment. Customs and Border Protection (CBP) raises several concerns about CAARS, as described under "operational risks and concerns," below. CBP, a component of the Department of Homeland Security, is a front-line agency whose officers perform such missions as operating border crossings and inspecting cargo entering the United States at seaports, airports, and land crossings.

Background

Two types of commercially available equipment for scanning cargo containers have been widely used since 2002.[34] One type, radiation portal monitors, passively detects radiation coming from a container. Another type, radiography equipment, creates x-ray-type images of a container. This type is more relevant to CAARS. Examples of the latter are the Rapiscan Classic Eagle and the SAIC VACIS.[35] Radiography equipment works as follows. A cargo container is driven between two components of the equipment, or the equipment moves over a stationary container. One component produces gamma rays from a highly radioactive substance like cobalt-60 or cesium-137, or x-rays from a linear accelerator. These photons emerge in a thin vertical fan-shaped beam and pass through the side of the container.[36] On the other side of the container, an array of detector material records photon intensity levels, which correlate to opacity, pixel by pixel. A computer assembles the pixels into a radiograph. A black or white spot indicates an area that is opaque to photons of the energy used. Key limitations of such systems are that they cannot differentiate between different types of material and cannot pick out threats from clutter. While a complete nuclear weapon might or might not be noticed, detection probability decreases as the threat object becomes smaller, so it would be difficult for current radiography systems to detect a small piece of HEU.[37]

A different physical principle enables a radiography system to search for materials with high atomic number (Z). Both the L-3 and SAIC CAARS systems utilize this principle. They use one or two linear accelerators to generate 6-MeV electrons and x-rays with energies from 0 to 6 MeV and, in the same manner, x-rays with energies from 0 to 9 MeV (abbreviated in this section as 6-MeV x-rays and 9-MeV x-rays). The former have the greatest number of photons at around 2 MeV; the latter, at about 3 MeV. (The third CAARS system utilizes a different principle.)

Photons of these two energies interact with matter differently. Six-MeV x-rays scatter when they strike electrons.[38] The amount of scattering is a function of the electron density of the material, so it increases with Z. As a result, high-Z material is more opaque to 6-MeV x-rays than is low-Z material, and creates a bright or dark spot on a radiograph. **Figure 8**, top panel, is a radiograph taken with 6-MeV x-rays. Nine-MeV x-rays interact more strongly with an atom's nucleus. When a nucleus absorbs a photon, it releases energy in the form of an electron-positron pair. This effect is proportional to Z squared. Thus high-Z material is much more opaque to 9-MeV x-rays (many fewer get through the material being interrogated because they are absorbed) than to 6-MeV x- rays.

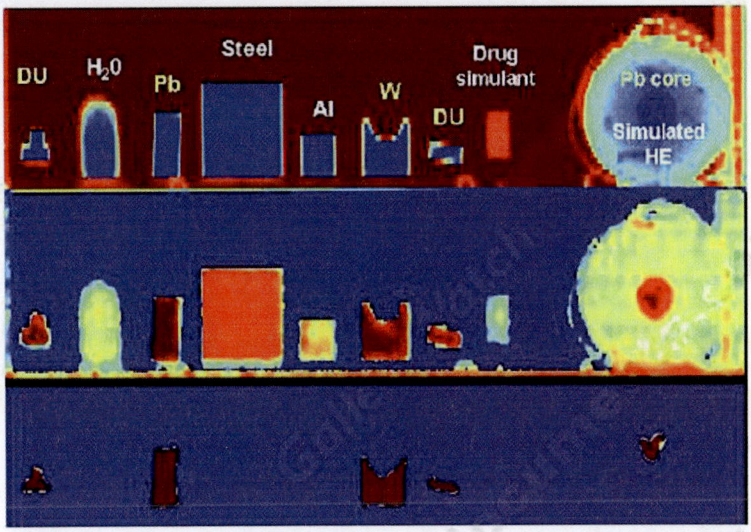

Source: These images were taken in January 2006 with an SAIC proof-of-concept laboratory prototype cargo imaging system called VACIS-Z: Detection of High-Z Material in Cargo. The VACIS-Z contract was funded by the Homeland Security Advanced Research Projects Agency of the Department of Homeland Security through a contract that predates DNDO (which was established in April 2005). The contract number is N41756-04-C-4200, awarded under Technical Support Working Group (TSWG) Broad Agency Announcement DAAD-03-T- 0024. For further information on TSWG, see http://www.tswg.gov/.

Notes: This graphic shows three radiographs of the same objects. (Abbreviations: DU, depleted uranium, i.e., natural uranium with most uranium-235 removed, used here as a surrogate for highly enriched uranium; Pb, lead; Al, aluminum; W, tungsten; HE, chemical high explosive) The top image was taken using bremsstrahlung photons with a maximum energy of 6 MeV. The middle image shows, pixel by pixel, the ratio of a 9-MeV and a 6-MeV radiograph. Higher-Z objects are more opaque to photons with energies up to 9 MeV than to those with energies up to 6 MeV, and pixels of such objects are represented in darker colors. The bottom image shows only the high-Z material from the middle image, which would automatically generate an alarm. In this image, the system would alarm on depleted uranium, lead, tungsten, and lead surrounded by a simulated chemical high explosive.

Figure 8. Dual-Energy Radiography

The L-3 and SAIC CAARS exploit this difference to create a so-called dual-energy radiograph, in which each pixel represents the *ratio* of opacity of that pixel to 9-MeV and 6-MeV x-rays. To create the radiograph, an algorithm assigns different colors to different ratios; in the middle panel of **Figure 8**, higher-ratio pixels are darker. Pixels with ratios above a certain value indicate high-Z material. While there is no absolute physical threshold between medium- and high-Z material, CAARS uses Z>72 as the threshold for high-Z material;[39] elements with Z>72 include tungsten, gold, lead, uranium, and plutonium.[40] Pixels of such material could be presented in a separate radiograph, as in the bottom panel of **Figure 8**. However, these data by themselves are not sufficient to trigger an alarm. A typical cargo container has many overlapping objects, so another algorithm is needed to separate them from each other. Still another algorithm utilizes the foregoing data to calculate the size and Z of individual objects, and on that basis determines whether to trigger an alarm.[41]

High-energy x-rays could also exploit a characteristic of U-235 and Pu-239 (as well as of U-238 and thorium-232, non-threat materials that are relatively uncommon in commerce): they fission when struck by photons with an energy above approximately 5.6 MeV. As discussed in the Appendix, the resulting fission products decay over many seconds, producing prompt and delayed neutrons and gamma rays. High-energy x-rays may thus be used to detect high-Z material in general and SNM in particular.

Technology description

This section focuses on the least complex and lowest-risk CAARS system, which is being developed by L-3 Communications Corporation. As **Figure 9** shows, it would use a concrete enclosure to minimize the radiation exclusion zone. The enclosure is 160 ft long in order to process two trucks at a time in 3 minutes so as to meet the CAARS throughput requirement of 40 containers per hour. The system would use one linear accelerator to generate 6-MeV electrons and (through the *bremsstrahlung* process) x-rays with energies from 0 to 6 MeV and, in the same manner, another accelerator to generate x-rays with energies from 0 to 9 MeV. A container would remain stationary as the beam is moved on a gantry over the truck. On the other side of the container are two detector arrays, one particularly sensitive to 6-MeV x-rays and the other to 9- MeV x-rays. Each array would record successive vertical slices of an opacity map of the container. The slices would be combined into a dual-energy radiograph as described above.

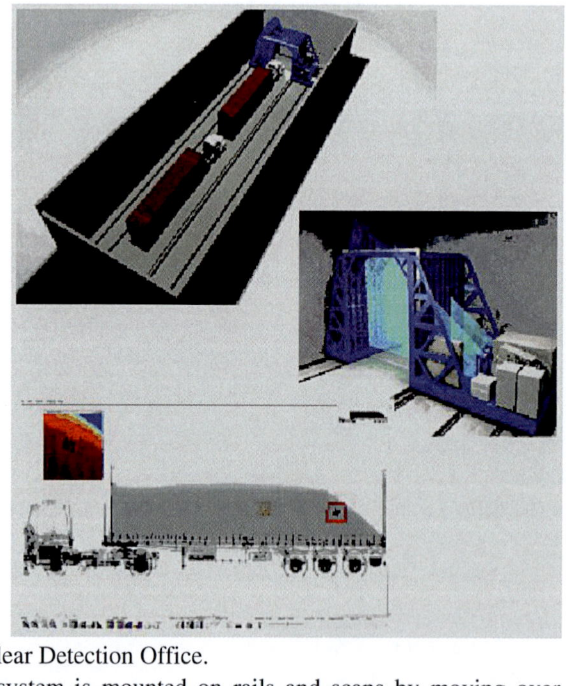

Source: Domestic Nuclear Detection Office.

Notes: The detection system is mounted on rails and scans by moving over the trucks. A concrete enclosure minimizes the radiation exclusion zone and thus the footprint. (middle) The unit has two accelerators (right) and two detector arrays (left). (bottom) This is a simulation of the image the system would return when it detects high-Z material in a cargo container.

Figure 9. L-3 CAARS Schematic Drawings

Potential benefits

If CAARS works as DNDO anticipates, it could automatically detect high-Z materials while providing standard radiographs so that it would have potentially little or no impact on CBP operations, such as efforts to identify "traditional" contraband (drugs, guns, explosives). As such, CAARS could, if successful, integrate the SNM detection mission with CBP's historical mission of detecting other illegal materials. DNDO anticipates that the automated detection of high-Z materials would not slow down the pace of screening, which could continue to be determined by the rate at which operators can examine radiographic images for detection of normal contraband, though CBP expresses concerns on this point, as discussed below. CAARS technology is being designed to scan at least 40 40-foot containers an hour.[42] DNDO anticipates that it would have a 90% probability of detecting 100 cc of high-Z material (such as a cube 4.6 cm, or 1.8 inches, on a side), and a false alarm probability less than 3 percent, both with 95% confidence.[43] DNDO does not count detection of high-Z material as a false alarm because such material may be used to shield SNM. Another possible benefit of the CAARS program is that it is developing novel technologies and advancing the state of the art; even if no CAARS systems were to be deployed, scientific and technical advances made through this program could be of value to future detection systems.

Status, schedule, and funding

In September 2008, Vayl Oxford, Director of DNDO, described developments with the CAARS program as follows:

> Consistent with any rigorous development and acquisition program, DNDO conducted system requirement reviews in November 2006 and preliminary design reviews in late May and June 2007 to assess the maturity of the CAARS technology. As a result, DNDO found that the technology was more difficult to implement than originally anticipated and determined that the technology should be demonstrated so that its full performance capability could be established prior to acquisition. It was also determined that the CAARS units, as currently designed, are too large and complex to be operationally effective. Finally, since 2006, there have been several technical advances in currently-deployed or soon-to-be-deployed NII systems that might provide some, but not all, of the desired capability. Accordingly, DNDO undertook a "course correction" in April 2008 and modified the three CAARS contracts to remove the "acquisition" component of the contracts, yet retain the demonstration and the test and evaluation (T&E) components of the contracts to allow collection of the required performance data.[44]

As of July 2008, the three CAARS contracts underway with DNDO have either completed or are quickly approaching the Critical Design Review milestone, the point at which the design has been completed and DNDO can give the contractor approval to begin fabrication of the prototype. Developmental test and evaluation (DT&E) for the three CAARS prototypes is scheduled to begin in mid-FY2009. This work would help develop the equipment and would provide data for the algorithm discriminating between high-Z and lower-Z material. Through this DT&E, the performance of the three prototypes would then be characterized by the start of FY2010.

Part of the course correction was establishment of the Joint Integrated Non-Intrusive Inspection (JINII) Program—"joint" because the project is a collaboration between DNDO and CBP, and "integrated" because it seeks to detect both traditional contraband and high-Z material. It is "non- intrusive inspection" (NII) in the sense CBP uses the term, namely, CBP personnel could clear containers, without physically having to open them, with high confidence that they do not contain contraband or SNM.

JINII has two main components. One is CAARS. The other is a test campaign that DNDO and CBP plan to begin in the spring of 2009 to examine the ability of existing, commercially- available radiography systems to detect high-Z material by means of operators visually inspecting radiographs. This is distinct from CAARS, which is intended to highlight suspicious areas automatically. In addition, systems have been developed outside of the CAARS program that have a limited capability—which DNDO anticipates would be less than that of CAARS—to detect high-Z objects in cargo automatically. The test campaign would eventually include such systems. DNDO states that if tests demonstrate this limited capability with commercial off-theshelf systems, deploying the latter systems could place capability in the field sooner than would be the case by using only CAARS systems.

Once the test campaigns are complete, DNDO expects to have data to quantify the performance and cost of CAARS systems and existing radiography systems so that it could perform a cost- benefit analysis that would consider both high-Z and traditional contraband detection. On this basis, it could recommend future development, operational testing, acquisition, or some combination, of the various systems. If DNDO, CBP, and Congress judge CAARS to be more cost-effective than existing systems, one or more systems with CAARS-level capability could be further developed or acquired through a competitive process yet to be determined.

A smaller part of JINII deals with contraband. While CAARS addresses automated detection of SNM, the agencies involved say that it is also important to improve the speed and accuracy of detection of contraband through development of algorithms; otherwise, the slower rate of detection of contraband would pace the processing of containers. Accordingly, DNDO, in coordination with CBP, has released solicitations to small businesses in FY2008 to address this issue, and plans to release further such solicitations to small businesses, national laboratories, and industry in FY2009.

The budget for the entire CAARS program is: FY2006, $16.3 million; FY2007, $26.4 million; FY2008, $31.8 million; and FY2009 requested, $26.1 million.

Risks and concerns

Scientific risks and concerns

The L-3 system is intended to be the least complex and lowest-risk CAARS system. The scientific risk to its hardware is low because it uses commercially-available accelerators and detectors. The software risk is potentially higher because algorithms to sort high-Z and low-Z materials on a radiograph automatically have not been fully developed and have only been modeled in a simulated environment, not tested in an actual operational one. The other two CAARS technologies are more complicated, so their scientific risk may be greater.

Engineering risks and concerns

The concept has been demonstrated in the laboratory, but it remains to be shown that it can be scaled up to the size CBP needs to scan containers. For example, can the algorithm to sort pixels into high-Z and lower-Z bins handle large enough quantities of data in a timely manner? At high enough energy levels, an accelerator can produce neutrons that require a large amount of shielding. The x-rays generated by the accelerator can also scatter from interactions with the cargo, also requiring shielding. A large footprint for shielding could preclude deployment of detection systems at some ports. At issue: Can radiation be kept low enough that the footprint does not become excessively large?

Cost and schedule risks and concerns

Slipping the CAARS schedule through the "course correction" to permit more time for R&D could make it easier to meet the new schedule, reducing schedule risk. More R&D could also reduce cost risk. On the other hand, scientific risk could result in cost and schedule risk.

Another cost and schedule risk is that work on three projects would reduce effort devoted to any one system, delaying each system (as compared to the schedule possible if the full funding had been applied to only one system) and increasing the cost of the total program. For example, it could be argued that a more efficient use of funds would be to focus only on the L-3 system because its development is furthest along. On the other hand, a multi-pronged approach may offer countervailing advantages. The simpler technology could (presumably) be deployed quickly, adding capability quickly and providing a hedge against failure of more advanced systems. More-capable systems could be deployed later, reducing the time pressure to develop them. Conversely, pursuing several approaches hedges against the prospect that the simpler system might encounter unanticipated problems that delay it to the point where it and a more advanced system could be deployed at about the same time. Whether these advantages justify the higher cost is always a matter of debate.

CBP notes that other companies are developing systems to detect high-Z material outside of the CAARS program, and argues that, because of the operational concerns discussed next, the money spent on CAARS could be better spent on such systems.[45] DNDO points out that the JINII program is developing and evaluating some of these systems.[46]

Operational risks and concerns

(1) A 3% false alarm rate, if achieved, would place a large burden on CBP CONOPS, as it would require responding to many false alarms. Alternatively, as discussed in the section on computer modeling, a high false alarm rate could lead operators to ignore alarms or to raise the threshold for alarms, increasing the likelihood of missing an actual threat. (2) Joel Rynes, Program Manager, CAARS Program, DNDO, states that an operator could clear some false alarms without intrusive inspection, such as by inspection of radiography images, reducing the rate of alarms requiring intrusive inspections to well below 3 percent. At issue is whether there is enough confidence in operator judgment with these methods to avoid intrusive inspections. (3) The system does not differentiate between SNM and other high-Z material; adding non-SNM high-Z alarms to false alarms boosts the non-SNM alarm rate and, presumably, the rate of alarms requiring intrusive inspections. Without a way to differentiate between SNM and other high-Z material, can CAARS meet its goal of having "little or no impact on CBP operations"? On the other hand, is any high-Z material suspicious and worthy of inspection, so that non-SNM high-Z alarms should not be counted as false alarms? (4) A CAARS program goal is automated detection of small amounts (100 cc) of high-Z material through 10 inches of steel, with a follow-on goal of penetrating 16 inches of steel.[47] Is that a

reasonable goal? If not, what is the maximum thickness of steel commonly found in cargo containers and what energy level of photons would be required to penetrate it? How would the added shielding required by equipment that generates photons greater than 9 MeV affect deployment at ports, where space is at a premium? Or would no practical energy level suffice to penetrate that much steel? Alternatively, would a simple "Bucky collimator," as described under "Background" in this section, make radiography much more effective? (5) The radiation produced by CAARS, or other high energy imaging systems, even at low levels, may cause interference with radiation portal monitors that have already been installed. It is important to minimize this interference, though it can readily be eliminated by turning off the accelerator when portal monitors are in use.

CBP raises other operational concerns. The first concern applies to the L-3 and AS&E CAARS systems; the others apply to all three. (1) Because of an apparent miscommunication between DNDO and CBP, DNDO thought that the dimensions of a one-of-a-kind CBP radiography system, 60 x 160 ft, were acceptable to CBP for large-scale acquisition, some 300 to 500 units nationwide. The L-3 and AS&E CAARS systems would use a concrete enclosure of that size (see **Figure 9**) for radiation containment. However, CBP states that no unit that size would fit in U.S. seaports, and that only four or five ports of entry on the U.S.-Mexican border could accommodate the units. (2) CBP expresses concerns that the radiation emitted by radiography equipment such as CAARS could require a large exclusion zone to protect workers. In contrast, radiation portal monitors emit no radiation, but passively sense radiation emitted by radioactive material. (3) CBP is concerned that total scan time would create an immense delay for container traffic entering the United States from seaports or land border crossings. While DNDO's goal is to have CAARS scan a container in 15 seconds, CBP notes that the time when the scanning unit is on is only a small part of the total time a scan requires. It points to the following sequence: a driver would pull a truck with a container into the scanning enclosure (depicted in **Figure 9**); the driver would leave the truck and go to a radiation-protected facility; the scanning equipment would move over the stationary container at a precisely-controlled speed; the truck and container would be scanned; and the driver would reenter the truck and drive it out of the enclosure. This sequence, CBP estimates, could take 5 minutes.[48]

Thomas Winkowski, Assistant Commissioner, Office of Field Operations, CBP, provided was emphatic on the need to minimize delays. Not referring specifically to CAARS, he said that rebooting and reinstalling software on some systems "can take seven or eight minutes. That's the kiss of death in my business from the standpoint of delays." Regarding CAARS, he said that CBP was "at the table" with DNDO in discussing requirements.

> But what our concern was was that the footprint was too big ... and the throughput, you know, for all your trucks to go through a car wash type system, as I call it, and the driver comes out, and you do your scan—that realistically presents a tremendous amount of problems from cycle time. So our position was that we really needed a different technology that was more flexible and that didn't have such a big footprint and require so much handling.[49]

Rynes stated that concerns such as the foregoing raised by CBP were among the reasons for the CAARS course correction and the establishment of the JINII program.[50]

Potential gains by increased funding

Rynes states that added funds could improve CAARS in several ways: (1) Further development could reduce the size of CAARS systems so they would be more readily deployable at ports. (2) Money spent (by CAARS or another project) to improve detector efficiency could also permit a less-powerful electron source to suffice, further reducing the requirements for shielding. (3) Added funds could improve algorithms to automate the processing of radiographs. (4) JINII is just beginning to develop algorithms to detect contraband automatically to increase the rate at which containers are inspected. Added funds would accelerate this development. (5) Added funds could expedite better integration and data fusion of CAARS with already-deployed radiation portal monitors.[51]

Potential synergisms and related applications

The preceding paragraph discussed synergisms by which technology developments could improve CAARS deployment characteristics or performance. Another synergism concerns the use of neutrons or high-energy photons to induce fission to discriminate SNM from other high-Z material. The current versions of CAARS do not include technology for this purpose. For example, they do not count neutrons or gamma rays that would be generated by photofission even though 9-MeV photons can cause that effect. DNDO issued a broad agency announcement in March 2008 to add this capability to CAARS and other high-energy radiography systems.[52] At the same time, such methods are under development for other projects; CAARS might draw on such work. Conversely, if CAARS included such discrimination capabilities, it could be competing with other such systems under development, possibly making one or more of them unnecessary.

SAIC CAARS: A Higher-Risk, Higher-Benefit Dual-Energy Radiography System[53]

The "problem" and "background" information for the L-3 system, as described above, are the same as for the SAIC system.

Technology description

Science Applications International Corporation (SAIC) is developing a system, as yet unnamed, as part of the CAARS program. The basic system would use an electron accelerator currently under development by Accuray Corporation.[54] The accelerator is "interlaced"—a single unit produces pulses of electron beams alternating between 6 and 9 MeV. Each pulse lasts 3 microseconds, with 2.5 milliseconds between pulses. These beams strike a copper target and generate photons with a spectrum of energies up to the highest energy of the electron beam. During a scan, the photons pass through a cargo container in a vertical fan-like beam. On the other side of the container is a detector array composed of multiple detector elements arranged in thin vertical bands. Each element records an opacity image of a narrow horizontal "stripe" of the container. An algorithm combines the stripes into a dual-energy radiograph, as described in the L-3 section. The system automatically flags for further inspection areas of high-Z material and areas with too much dense material for the photons to penetrate.

Figure 10 illustrates the configuration of a unit. A bridge-like structure is mounted on rails some 20 ft apart. A truck pulls a cargo container through the "bridge." In this figure, the accelerator is on the right side of the unit, with the detectors on the left. A steel plate 12 ft high by 20 ft long, with thickness ranging from 1 in to 7 in, is placed behind the detectors to stop photons that scatter off the cargo in the direction of the beam. This steel plate weighs some 70,000 lb but is part of the unit, so the unit is self-contained. A thinner plate but with the same length and width, on the right side of the image, is included to stop lower-energy photons that scatter backward from the cargo.

The key to the SAIC system is that the proprietary detector system operates at a photon flux (number of photons per unit time) one-hundredth that of conventional cargo imaging systems. A lower flux enables the accelerator to be much more compact. Accelerators with high photon flux require a high-Z material, typically tungsten, for the *bremsstrahlung* target because it can withstand the high heat from the electron beam. Such accelerators also use high-Z material to shield the photon beam. However, photons with energies greater than approximately 6 MeV produce photoneutrons (neutrons knocked off atoms by high-energy photons) when they strike high-Z materials, with the number of such neutrons increasing as energy increases. The heat generated by the beam with lower photon flux is low enough that copper can be used instead of tungsten. Copper is one of a few metals with a threshold greater than 9 MeV for producing photoneutrons. As a result, SAIC states, its system produces virtually no photoneutrons, eliminating the need for large concrete structures for neutron shielding. Further, since the system has a low beam flux, SAIC states that the need for shielding from x-rays that scatter in the cargo is greatly reduced.

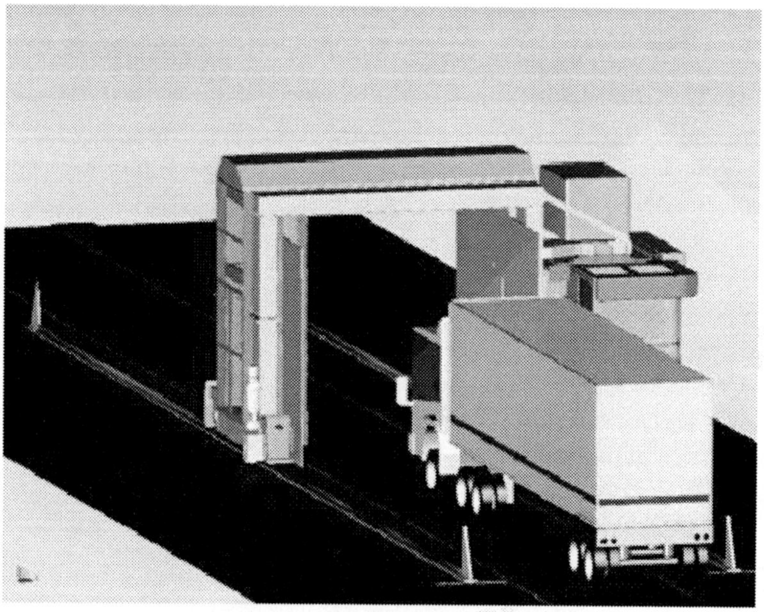

Source: SAIC.

Notes: The unit is mounted on rails some 20 ft apart. A truck pulls a cargo container through the "bridge." The accelerator is on the right side of the unit, with the detectors on the left. A steel plate 12 ft high by 20 ft long is placed behind the detectors to stop photons that scatter off the cargo.

Figure 10. SAIC CAARS Schematic Drawing

SAIC is considering an upgrade to its basic system in accord with DNDO's broad agency announcement of March 2008, and submitted a proposal to this effect in early 2008. The photon beam in its system is filtered.[55] After filtering, the beams of 6 and 9 MeV generate the greatest number of photons at about 2 and 3 MeV, respectively. The 6-MeV beam generates few photons energetic enough to induce fission in SNM, but the 9-MeV beam generates far more. Since fission generates gamma rays and neutrons, SAIC might add scintillating material to its system to detect prompt neutrons (those emitted with no time delay after fission) and delayed gamma rays,[56] and adding helium-3 tubes to detect delayed neutrons. This added capability would be intended to resolve high-Z alarms and reduce false alarms.

Potential benefits

Greatly reducing neutron flux offers several potential advantages. (1) Worker exposure to neutrons is reduced. (2) Lower flux requires less shielding, thereby reducing cost and footprint. (3) If the basic system is augmented by the capability to detect neutrons and gamma rays resulting from photofission, neutron detectors in this system would be able to detect neutrons when the beam is on because of the low neutron background. SAIC claims that its CAARS system is the only one with this capability. Since most neutrons released by fission of SNM are "prompt" (released immediately), the prompt neutron signal is much larger, and thus easier to detect, than that of delayed neutrons. Fission of SNM generates high-energy neutrons; by adding detectors that can discriminate between neutrons on the basis of their energies, SAIC expects that its system will be able to determine if a high-Z object is SNM. (4) The accelerator is very compact, 80 cm in length, also reducing footprint. (5) The system is designed to flag high-Z material automatically, reducing the burden on operators and the risk of operator error.

Status, schedule, and funding

SAIC is developing its basic system under contract to DNDO. Rex Richardson, Vice President and Principal Scientist, SAIC, stated in October 2008 that development and construction of that system are nearly complete; that all its performance requirements have been demonstrated using a high-fidelity laboratory prototype; and that that unit was able to find a 100-cc cube of high-Z material automatically through heavy shielding. SAIC has also completed development of key components of the basic system, such as the baseline algorithms and the photon detector system for radiography. Richardson stated that in September 2008 Accuray completed development of its dual-energy interlaced x-ray system and had its first successful full-power test of that system. In that and subsequent tests, he said, the system's performance exceeded SAIC's requirements. The program is being run as a preproduction program, with engineering completed, production drawings ready to use, and suppliers lined up. Richardson said in October 2008 that SAIC could take the first orders for production in March 2009 and could deliver the first unit of its basic system from a limited production run within six to eight months of receiving a production contract. SAIC estimates unit price at $5 million to $8 million. In September 2008, DNDO rejected SAIC's proposal for an advanced technology demonstration of the add-on capability to its CAARS system discussed earlier, to detect shielded SNM by detecting radiation released by fission of SNM induced by high-energy x-rays.

Risks and concerns

Scientific risks and concerns

This system requires completing the interlaced accelerator, yet its development is difficult. As a result, it involves more scientific risk than the L-3 system, which uses two separate accelerators, one for each energy level. An earlier attempt to develop an interlaced accelerator encountered a problem with the stability of the electron beam, with energies varying by a factor of two. The SAIC CAARS system requires that the Accuray accelerator demonstrate that it can repeatedly generate electrons of two energy levels, each within a narrow energy band. Richardson stated that the Accuray accelerator has demonstrated this capability, and that Accuray was conducting performance and endurance testing in October 2008, with formal acceptance tests for the system planned for November 2008. It remains to be seen if the accelerator has the endurance needed to operate in the field for months or years.

Another risk concerns the value of a possible CAARS add-on to detect radiation induced by photofission. Gamma rays so induced might be detectable, but neutron-absorbing shielding could hinder detection of neutrons. Jonathan Katz states, "Shielding against neutrons is very effective. No terrorist is going to ship an unshielded bomb. The whole business of detecting neutrons is a dead end."[57] While high-energy neutrons (14 MeV) might penetrate shielding materials and cause fission in SNM, the lower-energy neutrons produced by fission of SNM might not escape. SAIC states that the vast majority of cargo is low-Z, such as food and plastic, which would scatter and absorb neutrons, reducing the signal.

Engineering risks and concerns

SAIC has completed engineering drawings for the system and lined up its manufacturers. Yet development of the accelerator is not complete. At issue: If it does not perform as anticipated, how difficult would it be to modify the algorithms and reengineer the unit?

Another issue is whether enough helium-3 is available to fill the neutron detector tubes needed for large-scale deployment (see Appendix), though this would apply only to the version with the photofission add-on. Alternative neutron detectors are feasible, but if they are not part of the baseline for the add-on, any changes would require reengineering the system and, most likely, revising the neutron-detection algorithm.

A drawback of using an accelerator with lower photon flux is that the detector element must be larger, which degrades spatial resolution to 7 to 9 mm, vs. 3 to 5 mm for conventional systems. On the other hand, DNDO has set a standard of detecting 100 cc of SNM, e.g., a cube 4.6 cm on a side, so the significance of this loss of resolution is not clear for SNM detection, though it might affect the system's ability to detect other contraband.

Cost and schedule risks and concerns

CBP insists that scanning should interfere with the flow of commerce as little as possible. In response, DNDO requires the system to scan containers at a rate of 2.7 ft per second, the speed needed to scan a 40.5-ft container in 15 seconds. (This time refers to the actual time when a container is being scanned, as distinct from the requirement to process 40 containers per hour, which includes time for a container to move to and from the scanning equipment.) The 15-second requirement imposes a burden on the system. Improving the performance of the system at a given scan speed and photon output would require a significant redesign and additional cost. On the other hand, some ask, since radiographs of containers must still be scanned visually by operators to search for contraband, which may take more than 15 seconds, is a 15-second scan time needed? The system is to use an interlaced electron-beam source. Even if development of that source proves technically feasible, there is as yet no assurance that it will be available at the quantity, schedule, and cost needed to make the system competitive. (3) At a unit price of perhaps $5 million or more, some ask, is the system too costly to deploy in large numbers, and is the estimate of six to eight months from a production contract to delivery of the first unit of the basic system overly optimistic?

Operational risks and concerns

Uncertainties about the final configuration of the deployed system could affect operations. Space is at a premium in many U.S. and foreign ports. If a more powerful accelerator is needed, it could require more space, more shielding, and more standoff distance, and might increase concern among port workers about radiation exposure. If scan time of 15 seconds per container cannot be met, CONOPS would need to be changed. Either more inspection lanes would be needed, requiring more space and operators, or containers would be scanned at a slower rate, delaying the flow of commerce.

Potential gains by increased funding

SAIC states that with added funds it could (1) integrate detection of photofission with CAARS radiography, (2) hire more software engineers to speed development of algorithms to improve the system's ability to differentiate between materials; (3) pursue application of the system to detect contraband; and (4) work with DNDO and CBP to choose a platform (e.g., mobile road-going or tracked relocatable) for its system and develop a deployment plan.

Potential Synergisms and Related Applications

SAIC believes that there is significant potential for its dual-energy radiography technology to help detect contraband and explosives because it can differentiate between organic materials, which have a low Z, and most metals. The nanocomposite scintillator material being developed at Los Alamos might be of use to the SAIC system. Similarly, algorithms being developed for other gamma-ray detectors or radiography units might have elements similar to those being developed by SAIC.

AS&E CAARS: Using Backscattered X-Rays to Detect Dense Material[58]

This system addresses the same problem as the L-3 and SAIC CAARS.

Background

American Science and Engineering, Inc., (AS&E) is developing a CAARS system that would utilize a different physical principle than the L-3 and SAIC systems. Its core technology is "EZ-3D™," developed by Passport Systems, Inc. The term is an abbreviation for "effective" (i.e., average or approximate) Z (atomic number) of the material being detected, located in three dimensions. It is intended to exploit the principle that when x-rays are beamed at an object, the number of x-ray photons that scatter backwards (in the opposite direction from the beam) is strongly proportional to the object's Z.

When x-rays strike high-Z material, they knock pairs of electrons and positrons off atoms. (Positrons are electrons with a positive charge.) These electrons and positrons travel in all directions. When they strike other atoms, they create *bremsstrahlung* photons (x-rays) that also travel in all directions. By arranging photon detectors so that they detect x-rays scattered at a backwards angle (>90 degrees) from the beam, they could detect these x-rays and not x-rays from the beam. The number of these backscattered x-rays and their energy distribution (as shown in **Figure 11**) is an indicator of Z. In an ideal situation—for pure chemical elements, and with no intervening material—the number of x-rays is approximately proportional to Z to the fourth power, so that number is enormously higher for high-Z material than for medium-Z material. In the real world, most goods shipped (e.g., wood, steel, plastic, electronics) are not pure elements and differ in size and shape, and a container may hold mixed types of cargo. As a result, experiments have shown, the effect is somewhat less. The effect of Z on number and energy of backscattered x-rays is the key to EZ-3D; **Figure 11** shows, for five elements, the difference in backscattered spectra as Z increases.

This graph shows the number of backscattered photons (vertical scale) at each energy level (horizontal scale) recorded by a detector. The vertical scale is logarithmic, so vertical increments are larger than they appear. The plots are for five elements, from top to bottom, with Z: U, uranium, 92; Pb, lead, 82; Sn, tin, 50; Fe, iron, 26; Al, aluminum, 13. "Normalized 511" means that the graphs are "normalized" at 511 keV. That is, for each element, the number of counts at 511 keV is multiplied by the factor needed to set the count to a specified number (the same number is used for all elements in a scan) and counts at each other energy level for that element are multiplied by the same factor. This facilitates comparison across elements. For the region between roughly 600 keV and 1300 keV, the graph shows a difference between uranium and lead, and a larger difference between lead and tin. The distribution of points in the graph depends on the energy of the electron beam used to generate the x-rays via the *bremsstrahlung* process. For this graph, the beam has an energy of 2.8 MeV. Since almost all the x-rays have energies far below that level, with the peak number of x-ray photons at 300 keV, the number of counts diminishes at higher energy levels. For this graph, beyond about 1300-1800 keV, depending on the element, the number of background counts exceeds the number of backscattered counts, so that counts at and beyond those levels provide no useful data.

Technology description

Stephen Korbly, Director of Science at Passport Systems, Inc., describes the proposed design and operation as follows. A truck would pull a cargo container through the inspection unit at slow speed, 2.5 feet per second (1.7 mph). The container would first pass through a radiography unit to identify areas of dense cargo for the EZ-3D beam to examine in more detail.[59] The container next would pass through the EZ-3D unit. There, a Rhodotron[60] would generate a 9-MeV electron beam. This beam would travel through a series of electromagnets so as to steer the beam downward toward the container. The beam is designed to move back and forth transversely across the top of the cargo container to interrogate a "slice" of the container, as shown in **Figure 12**. If the system were to detect a volume of dense cargo, the beam could dwell on that volume longer to gather more data.

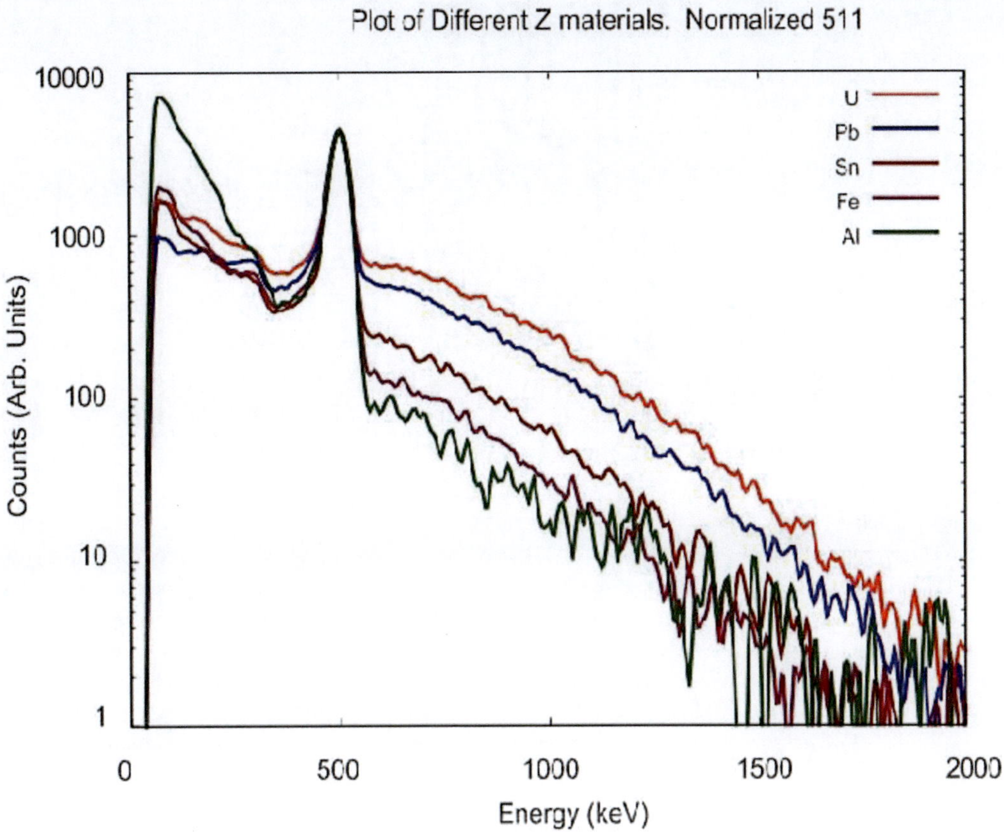

Source: Passport Systems, Inc.

Figure 11. EZ-3D™ Differentiates between Elements by Atomic Number (Z)

The electron beam would strike a water-cooled metal target, producing a spray of x-rays through the *bremsstrahlung* process. The x-rays would pass through a metal sheet that filters low-energy x-rays from the beam because they are of little value for detection but would increase radiation dose to the cargo. The remaining x-rays pass through a collimator, a slab of heavy metal with vertical holes drilled in it, so that only those x-ray photons traveling in the desired direction, downward, can pass through, forming a beam traveling in one direction. Collimation removes from the beam those x-rays traveling in other directions that would interfere with detection by scattering in the cargo or traveling directly from the x-ray generator to the detectors without going through the cargo.

The x-ray beam would pass downward through the cargo, generating other x-rays as described above, some of which scatter backwards. Very few other photons do so. Sodium iodide detectors are placed so as to detect these backscattered x-rays. The detectors are collimated so their field of view intersects with an x-ray beam.[61] **Figure 12** illustrates this configuration. The intersection of a beam and a detector's field of view forms a voxel. This intersection of two lines locates the voxel in two dimensions; the position of the "slice" of the container being examined locates the voxel in the third dimension. Each detector records the number of individual x-ray photons it detects in each voxel. AS&E expects the system to be able to scan a standard 40-foot cargo container for high-Z material in 15 seconds.

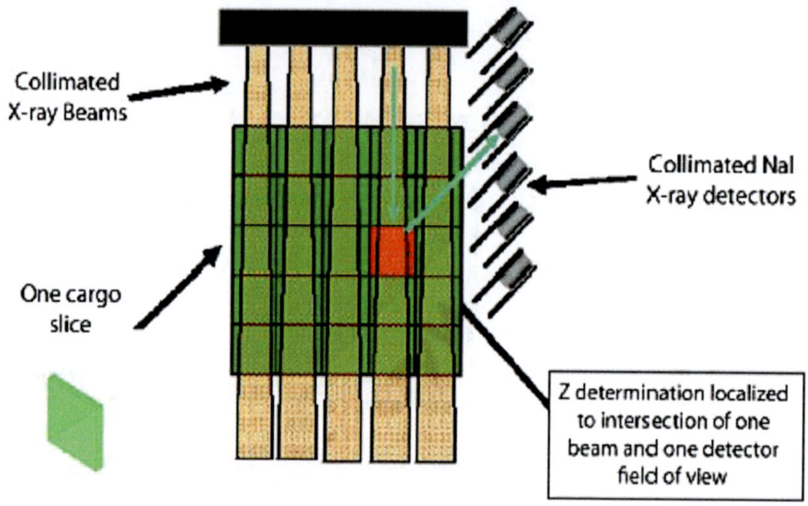

Source: Passport Systems, Inc.

Notes: This figure shows EZ-3D scanning one "slice" of a cargo container. The small diagram at the left depicts a slice, a transverse segment of a container. Many collimated x-ray beams exit from the collimator (bar at top of large diagram). They travel downward through the container. X-ray detectors are set to the side of the container, and collimated so that they view x-rays scattering at a backwards angle (i.e., greater than 90 degrees) from the direction of the beam. The intersection of an x-ray beam and the field of view of a detector forms a voxel whose effective Z is analyzed as shown in **Figure 11**. A beam and a field of view constitute two lines; their intersection locates the voxel in two dimensions. The position of the slice of cargo locates the voxel in the third dimension.

Figure 12. Schematic Diagram of EZ-3D Technique

Since the backscattered photons would pass through other cargo between the x-ray generator and a voxel, and between a voxel and a detector, the system is designed to account for, and subtract, the effects of this other cargo. Scanning a container produces radiography and backscatter data on how each voxel attenuates x-rays. An algorithm would integrate and analyze both types of data. In effect, to reconstruct the contents of a container, it would create a hypothesis about what is in the container where, and would calculate how closely the hypothesis matched the data. The algorithm would then alter the hypothesis iteratively until it provided a best fit with the data. The system is designed to alarm automatically when it detects voxels containing high-Z material and meeting other conditions. For example, the algorithm might alarm only if it detected a number of contiguous voxels of high-Z material so that it would not alarm each time it detected, say, a lead sinker. The process would repeat for each slice of the container. Korbly states that the algorithm has been shown to work in laboratory demonstrations.

Potential benefits

(1) Jeffrey Illig, the CAARS project manager at AS&E, stated in October 2008 that the AS&E CAARS would, if successful, look at 3-D voxels, rather than 2-D pixels; the latter approach, in effect, provides the average Z of an x-ray traveling through the entire width of a container as in the L-3 and SAIC CAARS systems. As a result, he says, the AS&E system is expected to generate more data by breaking the volume being inspected into smaller units,

simplifying the task of the detection algorithm. (2) Joel Rynes, Program Manager of DNDO's CAARS program, said that the EZ-3D effect is more specific to high-Z materials than is dual-energy radiography, potentially improving performance, though it will not be able to discriminate between (for example) lead and uranium within the scan time and dose to cargo limits required.[62] (3) The system is designed so that, if its radiography unit detects an area of dense material, it could direct the EZ-3D x-ray beam to spend more time scanning that area, further reducing the probability of a false positive or false negative. (4) Illig says that the system is designed to specify the location of a suspicious object in three dimensions; as a result, CBP personnel would know where to look in a secondary inspection, easing their task and reducing the time that a container is delayed for inspection. (5) If the system works as anticipated, it is claimed, it would automatically alarm on high-Z material, making it easy to use. (6) Because the x-ray beam is aimed downward, much of it would be absorbed by the ground, reducing the amount of x-radiation that escapes and reducing shielding requirements.

Status, schedule, and funding

AS&E has been working on its CAARS system since late 2006 under a contract that DNDO awarded for system development in September 2006. (The L-3 CAARS section provides the total cost of DNDO's three CAARS projects.) In early 2008, DNDO changed the goal of the CAARS program from system acquisition to an advanced technology demonstration (ATD). AS&E and Passport Systems have conducted numerous laboratory experiments at the University of California, Santa Barbara, to develop EZ-3D using the university's 5.3-MeV accelerator to generate x-rays to scan a 4 ft x 4 ft x 4 ft volume containing objects used to simulate commercial cargo. DNDO conducted a Critical Design Review (CDR) of the AS&E CAARS in October 2008, i.e., a review of the specifications for all components of the system and the links between them. DNDO approved the system's design at that time. As a result, the design is locked in and AS&E is able to begin to order hardware to assemble an ATD prototype system. Illig stated in early October that the system's design was complete and that, if the CDR is successful, AS&E would begin assembling a full-scale ATD system in November 2008. AS&E would also construct a shielded-room facility to develop the system in its anticipated production configuration. AS&E anticipates obtaining the first data from this ATD system in April 2009. After several months of developmental testing, AS&E would turn the prototype system over to DNDO so that agency could characterize and evaluate the system's performance using its own containers, cargo, and targets. If that phase is completed successfully, DNDO would decide whether to test the prototype under operational conditions, such as having CBP personnel operate it at a port of entry. Successful completion of that phase, in turn, would lead to a decision by DNDO on whether or not to purchase and deploy the system. Illig stated that AS&E could deliver the first production unit around the end of CY2010, with full-scale production commencing in CY2012; it projects the cost of its CAARS system at $8 million to $10 million per unit, assuming a buy of 25 units.[63] However, Rynes states:

> DNDO has no plans at present to purchase and deploy the CAARS systems. We will use prototypes of the three CAARS systems to collect data to feed a cost-benefit analysis that could lead to future procurements by DHS. The AS&E CAARS prototype will be a full-scale laboratory prototype that would still take substantial work to get it ready for a port

environment. It will take less work to get the SAIC and L-3 prototypes ready for a port deployment. The AS&E system has always been the high risk, high benefit solution.[64]

This report discusses this cost-benefit analysis under "status, schedule, and funding" in the L-3 CAARS section.

Risks and concerns

Scientific risks and concerns

(1) In theory, the system could differentiate between different high-Z materials, such as uranium vs. lead, if the materials were pure chemical elements and certain other conditions were ideal. In practice, so doing would take longer than differentiating between high- and medium-Z material, and impurities and interference from other cargo could make differentiation between different high-Z materials very difficult at best, so any high-Z material in a container could require a secondary inspection, delaying that container. (2) A container can include many types of cargo, and there is no requirement to declare the arrangement of the cargo. It would appear very difficult to develop an algorithm that can reliably eliminate x-rays scattered by cargo between the voxel being interrogated and a detector and reconstruct the locations of different chemical elements within a cargo container.

Engineering risks and concerns

(1) The system has not been demonstrated in a full-up configuration. AS&E, a Massachusetts company, uses an accelerator in California for its experiments, and these experiments use a lower-powered accelerator (5.3 MeV) than the Rhodotron (9 MeV). What risks, if any, are associated with shifting from a lower- to a higher-powered accelerator? (2) The integration of a linear accelerator for radiography and a Rhodotron for EZ-3D has not been demonstrated. (3) Passport Systems has simulated the performance of EZ-3D by gathering radiography data separately, bringing these data to the EZ-3D experimental facility in California, merging ("injecting") the radiography data into the EZ-3D data, and feeding the merged data into the cargo reconstruction algorithm. As with any simulation, one might ask how accurately the simulation reflects reality. (4) Will the AS&E system be able to meet the intended scan speed? There is always a risk of problems of this sort when one scales up from a laboratory demonstration system to an operational system. (5) Rhodotrons take a long time to build, and very few have been built. For example, AS&E has the world's 18th unit. Could the manufacturer, Ion Beam Applications (IBA), a Belgian company, build them faster and cheaper? IBA has told AS&E that if IBA gets a large order, it would build a facility in the United States to assemble Rhodotrons. Could it ramp up production in a new facility without major hurdles?

Cost and schedule risks and concerns

(1) If AS&E received a contract for multiple (e.g., 25) production units, could IBA ramp up its Rhodotron production facility on schedule? (2) The AS&E system is expected to be costly, and a main component of cost is the Rhodotron, at 2 million to 5.7 million apiece for single units.[65] Will IBA be able to reduce Rhodotron cost through research and through quantity production?

Operational risks and concerns

(1) Illig states that AS&E has extensive experience in designing systems and their user interfaces for CBP and other front-line users, but that AS&E has not consulted with CBP on the design of its CAARS system. Is that cause for concern, or is the user interface fairly standard by now so that extensive consultation is not needed prior to operational testing? (2) Illig states that the footprint of the unit is 60 ft by 160 ft mainly because it uses concrete walls for radiation shielding. The footprint is a concern for port operators because space is at a premium at ports; it is less of a concern at land border crossings. Illig states that the length could be reduced to perhaps 40-50 ft by 120 ft by using some means other than a truck to pull containers through the system; is that reduction sufficient? (3) The system generates radiation in two ways. A radiography unit uses a 6- MeV linear accelerator, aimed horizontally across a cargo container, to generate x-rays (through the *bremsstrahlung* process). Illig states that there is a "beam dump" to absorb x-rays on the other side of the container, and that the accelerator is pulsed, so that it is off most of the time, reducing the shielding needed. However, some x-rays scatter off detectors and cargo. The Rhodotron is on all the time, and will generate 9-MeV electrons, but is aimed downward so that the Earth absorbs most of the resulting x-rays. Further, the *bremsstrahlung* process generates x-rays in all directions; will the shielding be adequate? Accordingly, CBP and port personnel will be concerned about the amount of radiation that escapes. Will AS&E be able to provide satisfactory assurances on this point?

Potential gains by increased funding

Illig states that added funding would allow AS&E to build a test cell that had a Rhodotron and a linear accelerator for radiography together so that it could obtain actual data. That, he says, would permit engineers to develop algorithms to integrate both types of data using actual data. Added funds, he said, would also let AS&E purchase more test articles, conduct tests on more cargo configurations, and increase test time, all of which would characterize system performance better and reduce risk.

Potential synergisms and related applications

According to Illig, as part of the ATD for CAARS, AS&E would build a facility integrating backscatter and radiography technologies. This facility could provide a testbed for Passport Systems' nuclear resonance fluorescence technology and, more generally, for research into resolution of alarms caused by possible shielded nuclear material. Similarly, Rynes believes that the facility could have various applications:

> The AS&E CAARS prototype is being installed at MIT's Bates Linear Accelerator Center. The AS&E x-ray source (9 MeV, continuous wave) provides a capability that does not exist in the United States. After the CAARS program is complete, DNDO must dispose of all equipment procured under the contract. One option proposed is to keep the source at the Bates Center and establish the source as a "user facility" where researchers can come in to do experiments (e.g., NRF measurements at 9 MeV or experiments with prompt photofission). Another option is to let AS&E keep the source to use in a possible follow-on system (e.g., pilot deployment of a modified CAARS design). It is too soon to make this decision; it depends on how well the AS&E CAARS system performs in its upcoming tests.[66]

Muon Tomography[67]

The problem

Nuclear weapons or SNM may be hidden in cargo containers, automobiles, or elsewhere. Radiography might detect a fully-assembled nuclear weapon but could easily miss a small piece of HEU, and passive radiation detection is unlikely to detect even lightly-shielded HEU.[68] Use of neutron or high-energy gamma-ray beams holds the potential to detect SNM by itself or in complete weapons by inducing fission. But the radiation emitted by some such beams could require shielding and some standoff distance, possibly making it impractical where space is at a premium, and shielding of SNM might defeat this technique. Other properties of SNM are its high density and high Z. Such materials cause much greater deflection of muons, a subatomic particle, than do lower-density, lower-Z materials. Detection using muon tomography (MT) would not use a radiation source, avoiding concerns about radiation exposure or salvage fuzing. Yet while MT has been demonstrated in the laboratory, it remains to be seen if it can be converted to a system that will work in the real world. At issue: Is muon tomography an operationally feasible means of detecting nuclear weapons or SNM? Is it a cost-effective means of detecting high-Z material in vehicles with people inside them at land ports of entry and choke points?

Background

Muons are heavy subatomic particles generated when cosmic rays strike atoms in the Earth's upper atmosphere. Most muons travel at over 95% of the speed of light.[69] Given their speed and mass, they are highly energetic, with a mean energy of 3 billion electron volts.[70] As such, they are highly penetrating. For example, they can penetrate 1.3 m of lead or 15 m of water.[71] About 1 muon strikes each square centimeter of the Earth's surface per minute.[72] Matter deflects muons, with the degree of deflection determined statistically by the density and Z of the matter. As it happens, there is large separation between the average angle of deflection resulting from low-Z, low-density material in commerce, like food or plastic, and medium-density, medium-Z material like steel, and between the latter and high-Z, high-density material like tungsten, lead, or SNM.[73]

Tomography divides a solid object into many parts, determines a characteristic (e.g., density) of each part, and assembles the parts into an image of the object. A medical CAT (computed axial tomography) scan, for example, creates images of each slice (about 1 to 10 mm wide) of the body part being scanned. Muon tomography (MT) measures the trajectory of each individual muon before it enters a cargo container and again after it exits. In simplest terms, the intersection of the trajectories indicates the angle of deflection (and thus high, medium, or low density and Z) and the point of deflection, though the trajectory is more complex because a muon interacts with many atoms as it passes through a container. An algorithm integrates data from numerous muon trajectories to form a three-dimensional image of the container based on the density and Z of its contents. The statistical difference in deflection between high-Z, high-density material and other material, combined with muons' high penetrating power, is the basis for an MT system to detect SNM, whether in weapons or by itself.

Technology description

Decision Sciences Corporation (DSC) is developing an MT system, "Guardian MT," for use with vehicles and containers through a Cooperative Research and Development Agreement (CRADA) with Los Alamos National Laboratory (LANL). **Figure 13** is a schematic drawing of the system. It would work as follows. To determine the track of muons, the system uses "drift tubes," which are similar in shape to fluorescent light tubes. They are made of aluminum, with a wire running lengthwise through the center of each tube, and are filled with a gas mixture. The Guardian MT system uses arrays of drift tubes on the top, bottom, and sides of the object being examined to determine the incoming and exiting trajectories of individual muons. Each array consists of 12 cross-hatched layers of tubes, alternating between 2 layers in the "x" direction and 2 in the "y" direction. A positive charge is applied to the wires. When a muon strikes the gas in a drift tube, it creates a trail of free electrons that are drawn ("drift") to the wire by the positive charge at a known speed. These tubes measure the distance between the wire and a muon's closest approach to the wire. Two layers of "x" tubes establish a "y" measurement, and two layers of "y" tubes establish an "x" measurement. Combining data from each set of 4 tubes establishes a point on a muon's trajectory, and the system uses 3 points to define the trajectory.

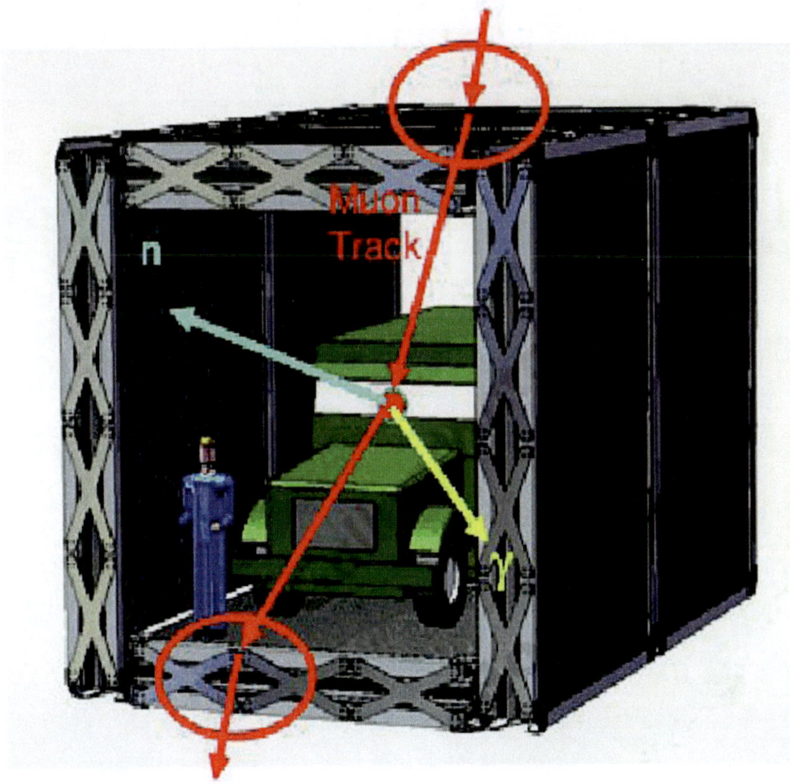

Source: Decision Sciences Corporation.

Notes: The truck remains in the station for about 20 seconds. The station detects scattering angle of individual muons, and counts neutrons (n) and gamma rays (γ). A muon track is shown entering from above the station and exiting below.

Figure 13. Schematic Drawing of Muon Tomography Inspection Station

For MT to work, the system must be able to identify each muon uniquely so it can match entry and exit tracks. According to DSC, the system's electronics can do this because the rate at which muons strike the drift tubes is far lower than the rate at which the electronics can process muon signals. Some 3,000 muons strike the drift tube array each second, and the electronics can read a muon hit in a millionth of a second. The odds of two muons striking a drift tube within a millionth of a second are 3000 out of 1,000,000, or 0.3 percent, so MT can uniquely identify a single muon 99.7% of the time.

At its simplest, if a muon interacted at only one point, the intersection of the two tracks would indicate the angle and point of deflection of an individual muon. The point of deflection would locate a voxel,[74] and the angle of deflection would indicate "scattering density," a combined measure of Z and density of the material in that voxel.[75] In practice, a muon interacts with all matter along its path, causing a displacement between its entry track and its anticipated exit track that provides information on the thickness of the material.

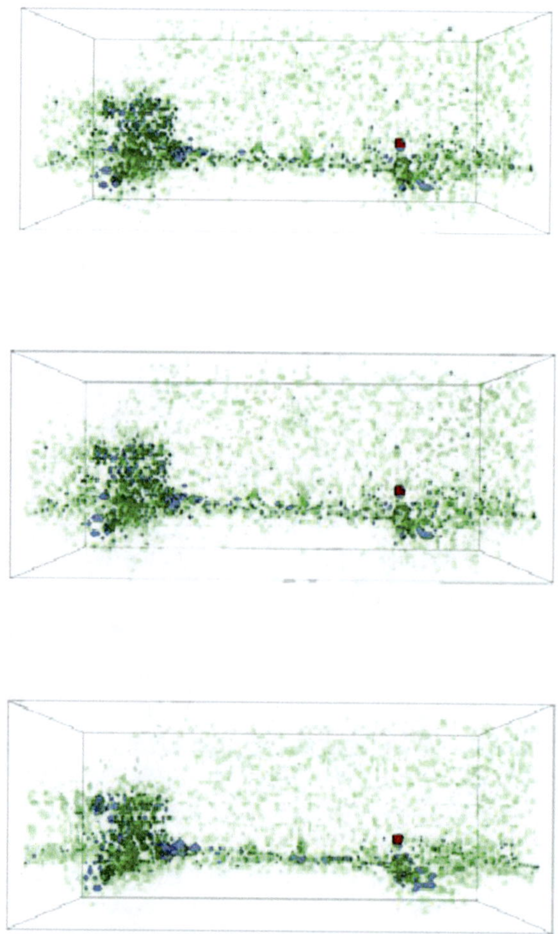

Source: Decision Sciences Corporation.

Notes: This figure shows scans of a truck at (top to bottom) 30, 60, and 90 seconds using simulated data. The dark spot "floating" at lower right (above rear axle) represents SNM and is highlighted in red.

Figure 14. Muon Tomography Resolution Increases with Scan Time

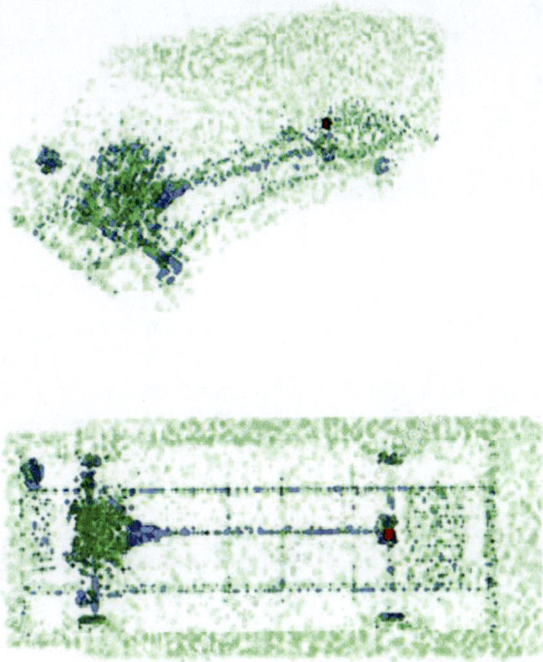

Source: Decision Sciences Corporation.

Notes: This figure shows scans at 90 seconds of side-angle and top views using simulated data. The dark spot above the rear axle represents SNM and is highlighted in red.

Figure 15. Muon Tomography Creates Three-Dimensional Images

From these data, an imaging algorithm calculates the degree of scattering of muons for each voxel and creates a 3-D image of the contents of the object being scanned. Resolution of the scan increases with time, as **Figure 14** shows, as each muon track pair adds data. The image is displayed on a computer screen and can be rotated so the viewer can visualize it as if in three dimensions, as **Figure 15** shows. According to one analysis, "only a few scattered muons are required to determine Z accurately enough to distinguish among the major groups of Z [high, medium, and low] with high confidence, and the value of Z is conveniently displayed as a color image."[76] Based on computer simulations and laboratory experiments using prototype equipment, DSC states that MT can differentiate between high-Z and medium-Z material, so it can pick out HEU hidden in a cargo of steel parts, or even hidden as a piston inside an engine.

DSC estimated in August 2008 that it would take less than 1 minute to clear most containers with 95% confidence and that Guardian MT could detect a cube of high-Z material 5 cm on a side in a container in that time. A longer scan time would increase confidence. This estimate is based on simulated data derived from physics calculations and on extrapolations from experiments conducted on laboratory-scale MT equipment, but not on experiments using full-sized MT equipment to scan loaded cargo containers enroute in commerce. A prototype detector that could be used to scan cluttered full-scale cargo scenes was in development and test at Los Alamos as of October 2008, and DSC states that as of that date the results had replicated the simulations and lab testing done previously. DSC plans to conduct testing in an operational environment using a field deployable unit in 2010.[77]

Based on computer modeling, DSC anticipates, on a preliminary basis, that MT will be able to distinguish between SNM and other high-Z materials. More scan time would be needed, perhaps 5-10 minutes to distinguish lead from HEU and twice that to distinguish tungsten from HEU. DSC plans to complete experiments before the end of 2008 on this capability and asserts that, if it works, it would be easier, less costly, less intrusive, and faster to scan a container for 20 minutes than to unload it. Christopher Morris, a physicist at Los Alamos who has done extensive research on MT, states,

> If muons were to identify a shielded container, I would advise taking as long as reasonable (perhaps ~1 hour) to survey a container with muon tomography before taking the risk of any invasive action. In this longer scanning time, one should be able to provide detailed images of the configuration of a threat object, estimate its yield if it is in a weapon configuration, distinguish between different high-Z materials radiographically, and carefully study the passive radiation signatures. This might avoid triggering a salvage-fuzed weapon.[78]

Detection might be made faster and more accurate by an algorithm that would subtract the known part of an image, making it easier to focus on suspicious objects. A library would be constructed with MT images of many models of trucks, cars, containers, and trailers. When a truck entered the MT system, a scanner would read its license plate. The algorithm would call up the MT image of that truck from the library, and would subtract that image, voxel by voxel, from the image of the truck generated by the MT system, leaving an image only of the cargo and any anomalous objects in the truck itself. This algorithm has been simulated but not tested.

To improve its ability to detect SNM, Guardian MT also detects neutrons and gamma rays. The drift tubes contain a small amount of He-3; neutrons react with this gas to produce a proton that makes an ionization signal that is readily identifiable, permitting the system to count neutrons. Absence of neutrons may help clear false positives, though hydrogenous cargo could absorb neutrons, preventing them from being detected; presence of neutrons would be suspicious. According to DSC, the system can also count gamma rays, though it cannot identify their spectra. When a gamma ray strikes the aluminum drift tube, it knocks electrons off the aluminum, which drift to the central wire and are recorded. According to DSC, the system can differentiate between electrons generated by gamma rays and by muons because the muon tracking algorithm can compute a track for muon-generated electrons but not for gamma-generated electrons, as the latter do not pass through a drift tube. A gamma-ray count above background levels would be suspicious, though there are many innocent gamma-ray emitters in commerce, while a low count would reduce suspicions.

Potential benefits

DSC claims that its MT system offers several benefits: (1) Because it uses naturally-occurring muons, it does not use a source to generate radiation, so it does not require shielding or a standoff distance to protect workers from radiation. The reduced footprint is of particular importance where space is at a premium, such as seaports. (2) Since it does not generate radiation, it cannot harm people or damage the contents of containers, so it could be used at land border crossings to search cars for SNM while the passengers are inside, speeding the flow of traffic. (3) Because it does not generate radiation, it cannot trigger salvage-fuzed weapons. (4) DSC claims that it is very unlikely to show a mass of high-Z material where none exists, or to fail to detect high-Z material. (5) Because muons are highly penetrating, they can be used to detect high-Z material even when shielded. (6) The detection algorithm detects, locates, and alarms for high-Z material automatically, greatly reducing the need for human interpretation. The display algorithm shows the operator where high-Z material is located and may provide information about its shape. (7) DSC claims that scan time is low. (8) DSC claims its system is highly reliable. "[W]e are essentially building a carbon copy of a prototype the lab [Los Alamos] has had in operation for several years. The system has proven to be extremely reliable both in place and in transit. The drift tubes have proven to be very reliable even when transported via air and surface and then reassembled."[79]

Status, schedule, and funding

In 1995-1996, LANL built and tested a small MT prototype (6 ft by 6 ft scan area) to demonstrate the detection of high-Z materials. Based on simulations and test data, DSC and LANL signed a CRADA in May 2007 to commercialize LANL's MT technology.[80] Since then, DSC has provided about $7 million to LANL as part of the agreement. LANL and DSC staff have built a larger scanner, 12 ft high by 12 ft wide by 16 ft long, large enough to fit a large SUV, that is being tested at LANL as of October 2008. Vehicles are to be tested in the device with various clutters as well as medium- and high-Z materials. As of October 2008, test results were expected by January 2009, a slip of three months from what was anticipated in July 2008. DSC estimates that it will cost another $1 million to complete the testing.

DSC provided the following update:

> DSC is developing a modular third generation technology platform suitable for scanning vehicle, cargo or up to tractor-trailer truck sized objects, and intends this to be used for operational testing in a port, airport or border crossing environment starting in 2010. Assuming that the system works and with expedited manufacturing, DSC estimates that pilot production units could be delivered in late 2010 or 2011. Units are to be built in 14-foot modules so they could be transported and configured to operational needs. DSC estimates that a unit 80 feet long, large enough to scan a tractor-trailer, would sell for several million dollars.

It anticipates that the system may be commercially available as early as 2011.[81]

Risks and concerns

Scientific risks and concerns

A concern is that terrorists might be able to counter MT. For example, they might try to smuggle uranium through a detector in pieces below the system's detection threshold, but that would risk exposing the plot to detection many times. Alternatively, since MT indicates the Z and density of a voxel, it might be possible to reduce those characteristics of HEU by forming it into pellets and mixing them with a low-Z substance. These techniques, however, would require fabricating HEU into a weapon-usable shape within the United States, providing clues to the plot. U.S. nuclear weapons use hollow pits,[82] in which a hollow shell of SNM is imploded to form a supercritical mass, initiating a nuclear explosion. Other nations might use this method as well. If terrorists were to obtain a state-made hollow-pit nuclear weapon, MT might not detect a pit of this type if the SNM shell were thin enough and voxels large enough that SNM did not fill most of a voxel.

Also at issue is the amount of time required for a scan. The Royal Society, the U.K. national academy of science, issued a report in March 2008 based on a meeting with dozens of experts from many nations, and stated,

> The key limiting factor is the time required for muon radiography, up to several hours to image only a cubic foot of a block of iron. According to a detector concept being developed by Los Alamos National Laboratory (LANL), it would take four minutes to image a cargo container. However, this would require detector panels perhaps the size of a large room. Moreover, once a shielded source has been identified it may take several hours to unpack the cargo to locate it.[83]

Several physicists who are working on MT disagreed with the scan time assessment and stated that increasing that time from 1 to perhaps 10-20 minutes may permit differentiation between SNM and other high-Z material, avoiding the need to unpack a cargo container. Experiments to be conducted with a prototype full-size MT scanner could help resolve this issue.

Robert Mayo, Program Manager, SNM Movement Detection/Radiation Sensors, and Advanced Materials Programs, Office of Nonproliferation R&D, NNSA, raised several other concerns:

> While muon tomography has been demonstrated to locate regions of interest in cargo that contains dense material, its material identification abilities are limited. What's more, quite large sized detection equipment, much more so than in other detection systems, is required, making MT rather impractical for many nonproliferation and national security applications. Most critically, however, MT is likely to require rather long scan times to adequately resolve dense images in cargo with a reasonable rate of false alarms. For these reasons, it is considered impractical as a screening technique. There are much more practical threat material identification and characterization tools being developed by various agencies of the federal government including passive spectroscopic and active detection technologies, as well as advanced radiography, all of which could be advanced and operationalized much more quickly with increased support.[84]

Engineering risks and concerns

Would it be more cost-effective for the system to focus on muon detection only and not on detecting neutrons and gamma rays? As noted in the Appendix, He-3 is in such short supply that the use of it for neutron detection in Guardian MT units does not appear feasible. Because of this issue, DSC began in June 2008 to look into other means of neutron detection, such as tubes coated with boron-10, though the neutron detection part of the system would have to be reengineered to accommodate new detectors, and algorithms would have to be revised, resulting in some delay. The net effect on cost depends on the material used and the cost of reengineering, but since He-3 is expensive a different neutron detection system might well lower unit cost. Another concern is that adding a separate unit to detect neutrons would increase the complexity of the system.

Cost and schedule risks and concerns

Is the schedule too optimistic? As of October 2008, DSC states that development is proceeding at a rapid pace. It envisions production after the third-generation prototype has been built and field- tested at an operational facility. It states that the detection algorithms are being modified into a form suitable for operational use. DSC plans a third-party test validation and life expectancy analyses and experiments. It remains to be seen if DSC can complete all these steps by the end of 2010 (the plan as of October 2008), and how a problem with any step would affect the schedule. Earlier, DSC had planned to complete these steps by the end of 2009. A cost issue is that, at perhaps several million dollars per unit, it would be costly to deploy enough MTs to scan all containers at seaports without impeding the flow of traffic.

Operational risks and concerns

It is uncertain at this time how alarms would be cleared. If MT is able to discriminate between SNM and other high-Z material, as its developers anticipate, then scanning a container for 20 minutes to resolve a high-Z alarm would be easier than unpacking a container. This anticipation, however, is based on simulations, not experiments. It may turn out that much longer scan times are required, in which case the balance between scanning and unpacking would be less clear. If experiments find that MT cannot discriminate between SNM and other high-Z material, then more intrusive means would be required. However, Katz estimates that these alarms would be few and could be cleared easily:

> I think the false positive rate would be a small fraction of a percent. There just aren't that many lead or tungsten ingots in commerce. When they are shipped they sit on pallets on the floor of a container (strapped down) and the rest of the container is empty. 30 tons of lead fills about 2.7 m^3 [cubic meters] out of the 64 m^3 container volume, so it is mostly empty space. A quick look is enough for secondary inspection.[85]

While this approach would reduce the burden of clearing high-Z alarms (for other systems as well as MT), the concern regarding nuclear smuggling is not clearing ingots of tungsten or lead shipped in this manner, but detecting SNM hidden in cargo. Another concern is how many ports could accommodate a system of this size.

Potential gains by increased funding

DSC states that added funds would enable it to improve the detectors and detection algorithms to provide more detailed imaging; incorporate sensors for additional signatures into the Guardian MT system to detect more threats and contraband; expedite the integration of gamma ray, neutron, and muon signals; develop the CONOPS that governs how the system would be used; and enhance and expedite engineering and manufacturing of the production version. Added funds, DSC says, would also support development of a library of MT images of different vehicle types; field trials for several types of applications, and development of an MT system to scan small boats, as described below.

Potential synergisms and other applications

The detection algorithm draws on those used for positron emission tomography, a medical technique to image bodily processes.[86] DSC is working with the positron emission tomography/single photon emission computed tomography imaging group at the University of California, Davis, to develop advanced imaging techniques. DSC states, "These algorithms will be applied to detection of SNM in our system and will feed back into the development of algorithms for medical imaging and lesion identification."[87]

MT may help detect SNM in other applications. The Coast Guard is concerned that small boats could enter a port carrying a nuclear bomb.[88] This is of particular concern for Miami, which small boats could reach from the Bahamas. Because muons penetrate about 15 meters of water, it might be possible to construct a muon detection system for boats, with an array of detectors above the water and several meters under water to detect entry and exit tracks of muons. Boats would stop in this array, perhaps for several minutes, while a tomographic image was built up. This application appears scientifically feasible, though engineering it would be considerably more difficult than for a land-based system. Another application might be detection of stowaways or contraband by locating suspicious voids in medium-Z material.

Scanning Cargo or Analyzing a Terrorist Nuclear Weapon with Nuclear Resonance Fluorescence[89]

Two problems

Nuclear resonance fluorescence (NRF), described below, seeks to detect SNM in containers. At issue: Is NRF a useful approach for this task?

NRF may address a second problem. Discovery of a nuclear weapon in a cargo container would require an urgent effort to disable it and to gather forensic data. Both efforts would benefit from detailed information about the weapon's design. Several techniques can provide such information. Radiography or MT can show the shape and location of components; discovering that the weapon had a thermonuclear stage, for example, would show that it was manufactured by a nation and could have much higher explosive yield than a terrorist-made bomb. Interrogation using neutrons or high-energy gamma rays can provide information about SNM. NRF may be able to provide different types of data. Knowing what kind of chemical explosive the weapon contained, or combining information on location of electronics with information on their chemical composition, or knowing the mix of isotopes and impurities in SNM, would aid in dismantlement or might point to the source of the weapon. At issue: Is NRF a useful approach to determining the materials in a weapon?

Background

When atoms of a given element are illuminated with photons above an energy threshold unique to that element, their electrons absorb the photons' energy and move to a higher energy level, a so- called "excited" state. The electrons then drop back to their normal state, emitting photons that are slightly less energetic than the inbound photons. For example, certain elements or minerals illuminated with ultraviolet light (which is more energetic than visible light but less so than gamma rays) give off visible light.[90] This emission of light is called fluorescence.

A different type of fluorescence provides more detailed information. Each isotope has a unique combination of numbers of protons and neutrons in its nucleus, so it vibrates at unique frequencies (the resonant frequencies). When the nucleus is struck by a photon at precisely that energy level—sometimes to within a few hundredths of an electron volt in a beam with photon energies spread over a range of 1 MeV or more—it will absorb the photon and move to an excited state. The nucleus then reverts to its initial state, giving off photons very slightly less energetic than those that it absorbed. This process is known as nuclear resonance fluorescence, or NRF. NRF produces a gamma-ray spectrum unique to each isotope (though different than the gamma- ray spectrum produced by radioactive decay). Identifying the spectrum of the emitted photons identifies the element and isotope. This is of particular importance in differentiating between fissile U-235, which can be made into a nuclear weapon, and non-fissile U-238, which cannot.

Unlike the use of neutrons or high-energy photons to stimulate the emission of neutrons or photons in SNM, NRF causes fluorescence in almost all isotopes of elements with Z>2 (helium), so it can identify a wide range of materials, not just SNM and other radioactive isotopes. Technical experts consulted for this report were aware of no other technology that permits identification of a weapon's materials without opening the weapon. CBP could also use NRF to identify other contraband and to check customs manifests.

Absorption of photons creates an additional signature. X-rays are generated in a broad spectrum of energies without sharp peaks. If a beam of such photons is sent through a cargo container or other object, and the detector on the other side can record the energy of each transmitted photon, a hole or "notch" in the spectrum at a certain energy level means that some particular material has absorbed photons at that energy level through NRF and then subsequently re-emitted the absorbed photons at about the same energy level. Since NRF photons are emitted in all directions, only a small fraction of them reach the detector. The energy level of the notch indicates what material is present. For example, the notch for U-235 occurs at 1.73 MeV.

Technology description

To detect NRF, an accelerator generates a beam of x-rays, a photon detector records the radiation spectrum generated by the material being interrogated, and an algorithm matches NRF peaks against a library of such peaks. Photons resulting from NRF are differentiated from the incoming photon beam because the latter produces a broad continuum of photon energies while photons generated by NRF produce very narrow peaks.[91] Further, the photon beam travels in a forward direction, while the NRF signal is emitted in all directions, so a photon detector placed behind and to the side of the material being interrogated (relative to the direction of the photon beam) detects photons traveling backward from the beam direction, which are mainly NRF photons. **Figure 16** illustrates this geometry.

DNDO is studying another approach to NRF using a beam of photons having a single energy level ("monoenergetic photons") near that needed to induce NRF in a particular isotope. At present, that method is technically difficult, costly, and requires large and delicate equipment, making it unwieldy for deployment in the field. However, according to DNDO, this method has been verified experimentally and the Stanford Linear Accelerator Center has demonstrated one type of accelerator technology "that might lead to a mono-energy photon source that could be compacted into a 20-foot cargo container."[92]

Passport Systems, Inc., is developing an NRF imaging system, Passport MAX (Material Advanced Inspection), under contract to DNDO to detect SNM in cargo containers. It uses a commercial electron accelerator with a beam that can be varied from 2 to 10 MeV, depending on the materials and containers being searched, to produce a photon beam with energies ranging from several hundred keV to the maximum energy of the electron beam. The beam is collimated;[93] as it scans a container, it excites nuclei in its path that emit photons. A germanium gamma-ray detector views the emitted photons scattered backwards from one region at a time, records their energies, and constructs a spectrum. The intersection of the collimated beam with the detector's view creates a voxel, and the spectrum shows the type and quantity of each isotope in that voxel. An algorithm constructs a three-dimensional image of the container's contents. Passport MAX would include other detector subsystems as well: EZ-3D, as described under AS&E CAARS; a radiography imager; and an NRF detector that detects notches in the transmitted photon spectrum. Passport Systems indicates that this approach could also be used to scan smaller items such as an aircraft cargo container[94] or a terrorist nuclear weapon.[95]

Because the NRF imaging component can examine only one region of interest at a time, the CONOPS envisions using the other components of Passport MAX to locate volumes of interest, and then using NRF to interrogate them. According to Passport Systems, "the complete system would scan a 40 ft container for SNM in an average of about 15 seconds. If there were indeed SNM in a container it may take longer (minutes) to identify the material as SNM. However, we anticipate that the actual number of containers with SNM would be very small."[96]

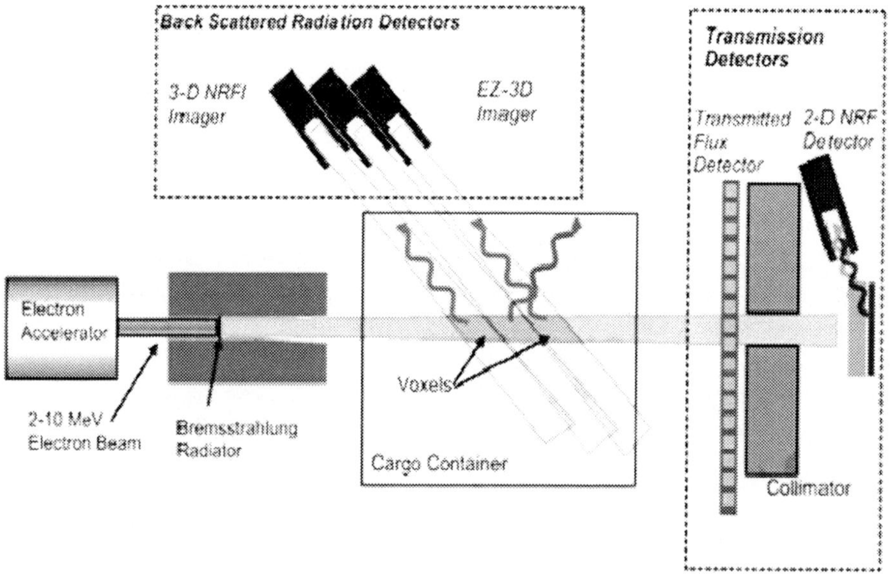

Source: Passport Systems, Inc.

Notes: From left to right: An accelerator generates a beam of electrons of 2 to 10 MeV. They strike a target (bremsstrahlung radiator), generating photons. A piece of high-Z metal allows only those photons moving toward the cargo container to exit. Photons pass through the container and induce nuclear resonance fluorescence, which releases photons of specific energies. NRF imagers "view" the container. The intersection of the photon beam and the "view" of each imager forms a voxel. Imagers are placed and aimed so that they view only photons moving backward from the direction of the beam in order to maximize the number of NRFinduced photons while minimizing the number of beam photons that reach the detectors. The system uses different types of detectors on the far side of the container.

Figure 16. Schematic Diagram of Nuclear Resonance Fluorescence System

Potential benefits

For detecting a terrorist nuclear weapon or SNM: (1) The system would identify each isotope, and would alarm on threat substances, with no operator input required. (2) While GADRAS must account for all the spectral data, the algorithm required for the Passport system need only account for spectral peaks, making for a simpler algorithm. (3) The system is to identify most isotopes that CBP finds in contraband, eliminating some false alarms that occur with radiation portal monitors, such as from radioactive potassium. (4) An NRF-based system can scan a cargo container quickly for SNM and high-Z shielding material. The average scan rate for the EZ-3D mode is 15 seconds, but the system would automatically adjust the speed at which individual containers are scanned. If it detected little attenuation of the beam, it would scan faster. Conversely, if EZ-3D identified a region of interest, the system would reposition the container to analyze that region using NRF and photofission signatures and would likely increase scan time for that container. The impact of such delays on average throughput rate would depend on such factors as CONOPS and number of anomalies in containers being scanned.[97] (5) Passport Systems states that laboratory experiments and simulations predict that Passport MAX will be able to meet the scanning requirement that DNDO has set for CAARS, i.e., that it would have a 90% probability of detecting 100 cc of high-Z material, and a false alarm probability less than 3%, both with 95% confidence.[98]

For characterizing a nuclear weapon: (1) This system could determine the composition of a nuclear weapon. These data would be of value for disabling a weapon and for nuclear forensics. (2) An NRF-based system can identify the isotopic composition of uranium or plutonium and any impurities, which would be of value in nuclear forensics. (3) A potential difficulty with radiation detection is that large quantities of shielding in a container may block photons or neutrons as they enter and leave the container. The shielding problem would diminish if this system were used against an already-identified terrorist nuclear weapon because the weapon would be shielded only by its casing. (4) A *bremsstrahlung* source generates photons over a wide range of energies, and detectors are able to record a similarly wide range of emitted photons. As a result, the system could identify multiple materials quickly.

Status, schedule, and funding

From 2004 to 2008, DHS awarded Passport Systems several contracts totaling $8.4 million to build a proof-of-concept (PoC) scanner that is fully integrated and functional. In 2005, the contract was transferred from the DHS Homeland Security Advanced Research Projects Agency to DNDO for management. According to Leon Feinstein of the Transformational and Applied Research Directorate of DNDO,

> The NRF PoC system demonstration and evaluation completed on August 4-6, 2008. The primary purpose of this PoC test is to demonstrate full functionality and automation. This requires all critical components to operate as specified in an integrated architecture similar to an operational scanner. The Passport PoC subscale system successfully demonstrated its ability to automatically select high-Z [regions of interest] with EZ-3D and auto-identify the isotopic content of [these regions] with NRF. This was accomplished with a variety of cargo and with a mixed set of high-Z material and contraband. Other NRF applications are still being explored including cargo manifest-checking and forensics.[99]

Most of the major hardware components that the system uses—accelerator, detector, computer, and display—are commercially available. Passport Systems estimates that its system will be available for commercial delivery by mid-2010 at a unit cost of $5 million to $10 million depending on system configuration. Feinstein states, however, that that system "will not have completed the DHS phased-milestones of development, testing, evaluation and cost-benefit analysis" by that time. He further states that DNDO is developing enabling technologies, such as improved accelerators and detectors, "that, if successful, could significantly reduce the overall size and cost [of the NRF system] (by more than a factor of two) as well as improve its performance and speed," though such technologies would not likely be ready for use in a commercial system by mid-2010.[100]

Many organizations have been conducting research into NRF to address national and homeland security issues since 2004. Lawrence Livermore National Laboratory (LLNL), Pacific Northwest National Laboratory (PNNL), and Passport Systems have collaborated to characterize the NRF response of U-235 and of Pu-239. PNNL has developed a computer model of NRF and has used simulations from it to reproduce results from laboratory measurements;[101] Los Alamos National Laboratory is developing another such model.[102] Duke University, Idaho National Laboratory, Idaho State University, LLNL, PNNL, Passport Systems, Purdue University, and University of California at Berkeley are conducting basic research on NRF. Passport Systems built its proof-of-concept prototype at University of California at Santa Barbara in order to use the accelerator in that university's free electron laser facility, and PNNL and Duke University conducted NRF measurements on U-238 and Pb-208 using the High Intensity Gamma Source at Duke as a source of nearly-single-energy photons.

Risks and concerns

Scientific risks and concerns

(1) A key scientific task is to conduct more experiments to identify the energy levels at which materials of interest undergo NRF. In particular, it would be useful to measure NRF spectra of isotopes of uranium and plutonium other than U-235 and Pu-239, and to measure spectra of materials used in nuclear weapons of other nations (e.g., for alloys) for purposes of nuclear forensics. (2) Another task is to develop the algorithms to identify materials quickly based on their NRF energies. (3) The Passport MAX geometry is designed to improve the signal-to-noise ratio by focusing on photons scattered backwards as a result of NRF, thus avoiding most of the photons generated by the interrogation beam, which travel in a forward direction. However, beam photons may have an energy range of nearly 10 MeV, while the energy range of photons that produce NRF in a particular isotope may be only a few hundredths of an electron volt. As a result, only one out of a few hundred million photons may have an energy level that produces NRF in that isotope, making it hard to detect NRF-generated photons. On the other hand, if the NRF system is tuned correctly, very few photons of the same energy as the NRF photons scatter backwards to the detector, facilitating detection by increasing signal-to-noise ratio. (4) While NRF has routinely been used to detect gram samples, the limitations of the detectable mass of various isotopes has apparently not been quantified, so it is not clear that NRF can be used to detect microgram (or smaller) quantities of all isotopes, possibly limiting the applicability of NRF to nuclear forensics.

Engineering risks and concerns

(1) More work is needed to develop the detection algorithms. (2) The accelerator for this system, called a Rhodotron,[103] is built to order, and it may take the manufacturer a year to build one;[104] unless the manufacturer can build them much faster, it would be difficult to procure Passport MAX units in quantity. (3) Once the tasks noted under "scientific risks and concerns" are completed, an engineering task would be to integrate the hardware and software into an operational system, which will require many tradeoffs between cost, performance, and schedule.

This may be particularly complicated for Passport MAX given the many separate detector units it uses, as shown in **Figure 16**.

Cost and schedule risks and concerns

The projected unit cost, $5 million to $10 million, is at a level that might preclude ordering large numbers of units. Regarding schedule, it is difficult to predict when an early-stage development program would become commercially available given the work that must be done and the possibility of unanticipated problems. As of August 2008, DNDO is funding enabling technology that may, if successful, significantly reduce the cost of the commercial system, reduce its overall size, and increase its NRF performance speed.

Operational risks and concerns

The system would need to be large enough to hold a tractor-trailer truck. The Rhodotron used in the Passport MAX generates a considerable amount of radiation that requires containment. To meet these requirements, the system is completely enclosed, and a preliminary estimate by Passport is that it would be 90 to 100 ft long; 20 to 30 ft wide at its widest point, where the detection equipment is located; and several stories high at that point. These dimensions are dictated by the requirement to scan a container that is 40 ft long, 9 ft wide and 14.5 ft high. A system for smaller objects would be correspondingly smaller. Passport states that the enclosure will reduce the radiation dose outside the system to levels low enough to not require an exclusion zone. On the other hand, the system is quite large using existing commercial off-the-shelf technologies, which could be a problem in ports where space is at a premium.

Potential gains by increased funding

Passport Systems, a small company, does not have some key equipment of its own; for example, it is using an accelerator at the University of California at Santa Barbara as the photon source for its prototype. Added funds, Passport says, would enable it to buy its own equipment; order longleadtime items, such as Rhodotrons and detectors; and hire more staff for engineering, algorithm development, and manufacturing. The prototype Passport MAX uses a considerable amount of expensive off-the-shelf hardware; with added funds, Passport says it could design less costly components specifically for use in its system.

Potential synergisms and related applications

(1) NRF could help identify illicit cargo in addition to SNM, such as explosives or chemical weapons. (2) NRF could help verify the manifest (list of contents) of a cargo container. (3) NRF might be used in nuclear forensics to identify rapidly the materials present

in radioactive debris from the detonation of a terrorist nuclear weapon. (4) In nuclear nonproliferation applications, it could be used to analyze the isotopic composition of spent nuclear fuel. (5) New detector material could improve the sensitivity and resolution of the system, reducing the amount of electrical energy needed to run it and the amount of shielding needed, thereby reducing acquisition and operational costs.

Detecting SNM at a Distance[105]

The problem

It is hard enough to detect SNM at a range of several meters. Yet the ability to detect SNM or nuclear weapons at a kilometer or more, or in territory to which access is denied, would have great value in the fight against nuclear terrorism, such as locating SNM along a smuggling route, in a distant vehicle or facility, or in ships at sea, or finding a terrorist nuclear bomb in a city. This capability would also help protect military forces by enabling detection of terrorist attempts to use a nuclear weapon to destroy a military base or a carrier battle group. Making the task harder still, SNM might be shielded, and these missions would require a high search rate because of the need to scan a large area quickly or because SNM might be visible to the system for only a brief time. A system to perform this mission would be of value. At issue: Can a system be designed and deployed to detect SNM at an operationally useful standoff distance and search rate?

Background

The Defense Threat Reduction Agency (DTRA), a unit of DOD, is pursuing capabilities to increase the distance at which SNM can be detected, with the goal of increasing it to 1 km or more. DTRA states its rationale as follows:

> DTRA/DoD's motivation for pursuing Active Interrogation arises from the context in which a search for nuclear material would be conducted—potentially in a hostile environment over large areas (due to limited intelligence pinpointing the exact location of the material). DTRA/DoD is also challenged by the need to make the deployed equipment robust/well engineered enough to survive a range of harsh environments which is different from equipment use in a static mode at a border checkpoint.[106]

A scientific panel said, "Radiation attenuation due to shielding is an exponential process and so even moderate amounts of shielding can have significant effects. At 10 metres, the radiation emissions of shielded gamma ray and neutron sources are at, or below, natural background rates in almost all cases."[107] DTRA states, "The only way to overcome this physical reality is to stimulate the radiation emitted by SNM to a level many times above background. This can be done, for example, by using a beam of high energy photons to artificially induce photofission, and then detecting the resulting fission signatures. Beams of other types of radiation also have the potential to increase these detectable signatures through other reactions with the SNM."[108]

Technology description

DTRA is sponsoring several remote-detection systems. This section considers only the one closest to deployment, the Photonuclear[109] Inspection and Threat Assessment System (PITAS), which is being developed by Idaho National Laboratory (INL). **Figure 17** shows schematic and cutaway views. The main hardware components are an accelerator and detectors. PITAS would use a powerful linear accelerator, capable of generating a 30-MeV electron beam, to create x-rays that would be aimed at the target. These x-rays have a range of energies, with the maximum equal to that of the electron beam. Air attenuates lower-energy x-rays in the beam, so the target would be struck by x-rays well above the energy needed to induce photofission. Detectors might be located in the same unit as the accelerator. However, radiation spreads out (and diminishes in intensity) as the square of the distance, and the atmosphere would absorb many neutrons and gamma rays, so detection can be more effective if the detectors are separated from the accelerator and placed near the target. For example, detectors might be placed next to a smuggling route or on unmanned aerial vehicles with the accelerator some distance away. The x-ray beam would be used in a pulsed mode with the detectors attempting to detect SNM signatures, such as delayed neutrons and gamma rays (see Appendix), in the intervals when the beam is off.[110]

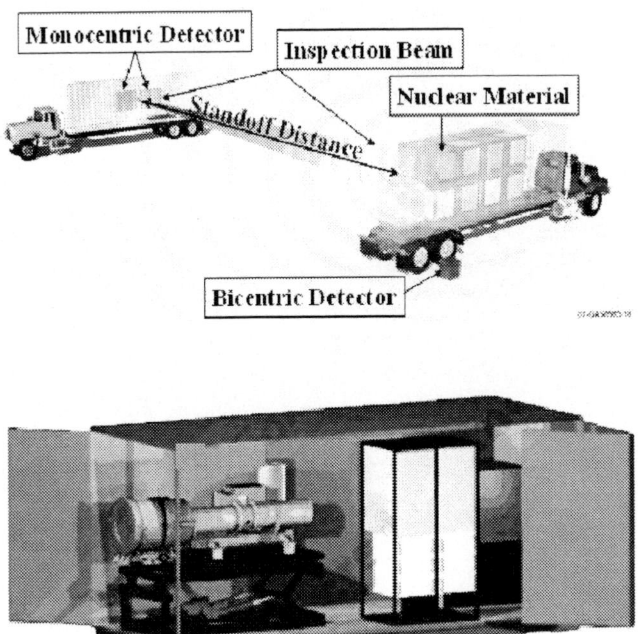

Source: Defense Threat Reduction Agency.

Notes: The top image shows the concept of operations for PITAS. A "monocentric" detector is one that is located with the PITAS unit; a "bicentric" detector is one that is located elsewhere, typically near the object being inspected. The bottom image is a cutaway view. The accelerator is on the left; it generates bremsstrahlung photons. The refrigerator-shaped object is the power supply and associated data collection systems, and the object on the far right is the power converter and timing circuits. The detector unit is not shown but could be included in the container housing the PITAS unit.

Figure 17. Schematic Diagram of Photonuclear Inspection and Threat Assessment System (PITAS)

Potential benefits

If PITAS works as anticipated, it would enable the United States to detect SNM at standoff range for the first time. This capability could be used for counterproliferation, counterterrorism, and force protection.

Status, schedule, and funding

DTRA declined to provide information on status, schedule, and funding on an unclassified basis. DTRA does not have a unit cost estimate but, compared to cargo inspection systems, unit cost is arguably much less relevant to total system affordability because the system would presumably be ordered in small quantities and quite possibly used for military missions of critical national importance.

Risks and concerns

Scientific risks and concerns

By one calculation, even a collimated x-ray beam could not place enough x-ray photons on a small piece of SNM in the target (such as a cargo container) to induce a detectable set of unique signals under ideal conditions.[111] Trying to detect SNM at a distance under real-world conditions, such as a container being towed by a truck along a winding road, would be even more difficult. How effective would shielding be at stopping inbound x-rays and outbound neutrons and gamma rays resulting from fission? (3) Gamma rays and neutrons resulting from fission of SNM would radiate in all directions, spreading out with distance. A source that caused 1,000 photons per second to strike a detector 1 meter square at a distance of 1 meter would cause only 1 photon to strike that same detector every 1,000 seconds at a distance of 1 km—disregarding attenuation by the air.[112] Yet enough photons must strike the detector to be detected and to be differentiated from background sources.[113] Regarding neutrons, fission generates many fewer neutrons than gamma rays, and the low-Z atoms in the air would attenuate neutrons strongly. Would enough gamma rays and neutrons reach the detectors to be detected? Another concern is that very little research has been done on remote detection. As a result, the developers of PITAS would have to conduct field tests to determine what signals the beam generates in SNM and how to detect them, and would have to develop algorithms to process the data. DTRA states that it is performing laboratory and field tests of candidate systems as well as calculations to address these issues.[114]

Engineering risks and concerns

One candidate neutron detector for PITAS is helium-3 drift tubes, which have a thin metal wall and a wire stretched lengthwise down the middle of the tube. Would these tubes be rugged enough to deploy as part of a mobile system that might operate in difficult terrain and extremes of weather and temperature? More generally, can the entire system be made compact enough to operate under such conditions, yet sensitive enough to operate effectively?

Cost and schedule risks and concerns

DTRA declined to provide information on this topic on an unclassified basis.

Operational risks and concerns

The accelerator would generate beams of very high energy, and might not be heavily shielded in many mobile applications. It thus poses some radiation risk to the operator. Could radiation exposure be made low enough? Whether the risk is deemed acceptable would depend on exposure, which in turn depends on radiation output, frequency of use, and shielding, and on the urgency of the mission. Regarding the latter point, detecting SNM that intelligence indicated was being smuggled along a known route for sale to terrorists would justify more exposure risk than would monitoring that route on a routine basis. Further development of accelerators might reduce the shielding required.

Potential gains by increased funding

DTRA estimates that doubling funding for PITAS would shorten time to deployment by perhaps a year.

Beyond PITAS, DTRA is contemplating another remote-detection system for which, it asserts, added funds would shorten the schedule by several years. That project would use a proton beam with very high energies. DTRA regards it as very promising because, it believes, these protons may be able to defeat shielding at 1 km. DTRA maintains that this project would be paced by investment. DTRA states that since most of the technical risk is at the start of the project, more funds would permit exploration of many paths simultaneously so that potential pitfalls and opportunities could be identified at an early stage.

Potential synergisms and related applications

This system could be used in support of the Proliferation Support Initiative, which seeks to stop shipments of nuclear weapons and other weapons of mass destruction worldwide. For example, PITAS could help search a ship suspected of carrying nuclear weapons, SNM, or uranium ore. PITAS might benefit from nanocomposite scintillator material. PITAS requires a material with substantial (but not extreme) resolution that is rugged enough to deploy under harsh conditions and inexpensive enough to deploy a large panel in order to capture more of the signal.

3. OBSERVATIONS

Observations on Progress in Detection Technology

Equipment commercially available at the time of the 9/11 attacks was limited in its capability. PVT radiation detectors could detect radiation but could not identify isotopes, and shielding SNM might defeat detection. Radiographic equipment could reveal dense objects, but relied on operator skill to flag potential threats. It might be possible to hide a nuclear artillery shell in a cargo of dense objects, and it would be difficult to pick out a small piece of SNM. Resolving alarms required time-consuming methods, such as using hand-held radioisotope identification devices or unpacking a container.

Capabilities of existing systems can be improved incrementally, such as by using different detector material, computers, algorithms, or CONOPS (e.g., scan time).

Systems now under development have the potential to reduce false positives (speeding the flow of commerce) and false negatives (improving security). Fission that neutrons or x-rays induce in SNM generates unambiguous signals. Dual-energy radiography detects high-Z material automatically. EZ-3D reveals high-Z material hidden in medium-Z material, and might be able to differentiate SNM from other high-Z material. These approaches detect useful signatures, but have drawbacks as well, such as low signal strength, complexity, high cost, or large size. The task is to utilize these signatures and minimize drawbacks in a system that can be fielded. Other technologies, such as improved detector material and improved algorithms, also have the potential to improve detection capability.

Detection technology appears to be advancing faster than many have expected. In March 2008, the Royal Society issued a report on nuclear detection that resulted from a workshop of 70 experts from around the world. The report found, "In the medium term (5-10 years), there are promising opportunities to develop new technologies, such as muon detection systems. In the long term (10-20 years) detection could benefit from advances in nanotechnology and organic semiconductors."[115] Yet a muon tomography system may be commercially available as early as 2011, and some expect nanocomposite scintillator material to be available for transfer to industry by September 2009. Whether these schedules are realistic remains to be seen. Similarly, Leon Feinstein states, "Advances in accelerator technology over the past 2 years may also speed up the availability of single-energy, laser-like photon beams with energies of interest to NRF and photo- fission, increasing stand-off ranges and expanding low-dose application opportunities."[116]

It is easier, less costly, and potentially more effective to accelerate a program in R&D than in production. DTRA believes that a significant increase in funding for proton beam technology, a standoff detection technology in early R&D, might shorten time to deployment by several years by enabling researchers to consider many technical alternatives simultaneously to determine the most promising approach faster. It is hard to attain large schedule gains by accelerating production; such gains may entail high cost, such as multiple shifts or more production lines; and a rush to production may cause a project to fail. While R&D projects may also fail, more risk is tolerable in R&D because the investment is much less.

A modest amount of money spent in R&D can avoid looming problems. For example, GADRAS, a widely-used algorithm for detecting SNM and other materials, runs on the standard Microsoft Windows operating system (OS) for personal computers. Microsoft introduces new generations of OSs from time to time. Typically, new OSs will support programs written for several generations of previous such systems. However, the Graphic User Interface (GUI) for GADRAS is written with the Visual Basic 6.0 compiler, which Microsoft no longer supports. At some point, Microsoft will likely introduce a new OS that will no longer support applications that are written with this compiler; GADRAS would then be unavailable to its users until it is updated. According to Dean Mitchell, who created GADRAS, updating that algorithm to run on current- generation OSs would avoid that problem, at a cost of perhaps $1 million a year for two years.[117]

R&D that leads to products that many systems can use may have a large impact on detection capability. Many detector systems have common elements—an accelerator, gamma-ray detector material, computers, algorithms—so improving any of these "building blocks" might improve the capability of many detector systems, including those in the field. Improved gamma- ray detector material can improve sensitivity, reduce cost, or both. An improved algorithm can boost performance. A more powerful computer permits the use of a more powerful algorithm, which may reduce false positives and false negatives.

On the other hand, it may not be possible to upgrade systems simply by swapping new components for old. Edward McKigney listed possible difficulties in the (hypothetical) case of upgrading systems by substituting higher- for lower-performance detector material:

> (1) Detector modules that cannot detect light with high efficiency would need to be redesigned. This is particularly relevant for existing portal monitors that use plastic scintillator material, where the optical design is poor. (2) Electronics for converting signals from detectors into data for algorithms ("readout electronics") that are not suitable for high-resolution readout and analysis, or are mismatched for the technology (such as if the old electronics read electrical charge while the new ones read optical signals), would have to be replaced. (3) Data analysis algorithms that cannot process signals from the new detector module would have to be replaced. (4) The volume of data from the new detector module might be greater than the existing algorithms, data transmission system or computers could handle, requiring new computers, algorithms, data transmission system, or some combination. (5) Electrical power systems would have to be changed if the power requirements for the old and upgraded systems did not match.
>
> So, at the extremes, it might be possible to upgrade only the detector module, or the only features of the old system that would remain after an upgrade would be the wide spot in the road and the guard shack. I would recommend that **the next generation of detector systems should be more modular** so upgrades could be done while retaining as much of the value of deployed systems as possible.[118]

Synergisms may arise between technologies. Beams of neutrons or high-energy gamma rays used to induce fission in SNM may harm some products, expose stowaways to high doses of radiation, and require shielding to protect workers. Improved detector material and algorithms could lower the amount of radiation required for this technique, perhaps making it more usable for scanning containers.

Technical advances can place two systems in competition unexpectedly. Work is underway on several systems designed to induce fission as a way of detecting SNM. CAARS was not begun as a system of this sort. However, DNDO is investigating a technology add-on to give it that capability. If work proceeds on that path, CAARS and other such systems could be in competition.

Competition at the R&D level may be desirable. William Hagan, Assistant Director, Transformational and Applied Research, DNDO, states,

> if we can squeeze additional functionality out of a system, we want to do that. This will cause various approaches to be in competition for achieving a capability at the R&D stage but that is what we want to do so we can drive towards the most effective.

> More generally, I think that having multiple organizations pursuing the same R&D goal is a good thing because it allows for different approaches or more capable organizations to compete for the objective. This is a very effective mechanism in R&D. A classic recent example is the race to decode the human genome. Another is the race for commercial space flight. This kind of competition goes on all the time in the basic research community and I think we should encourage it. There is, of course, some limit to this, but we are far from that limit right now for radiation detection.[119]

"Concept of operations" (CONOPS) is crucial to the effectiveness of detection systems. CONOPS details how a detection system would be operated to gather data and how the data would be used. Without it, a detection system would be valueless. Since CONOPS and systems are mutually dependent, the design of each must take the capabilities and limitations of the other into account.

Current equipment to detect and identify SNM makes use of two main signatures of this material, opacity for radiography to detect SNM, and gamma-ray emissions for spectroscopy to detect and identify SNM. However, as discussed in Chapter 2 and the Appendix, **SNM has many signatures in addition to opacity and gamma-ray emissions, and some systems under development attempt to make use of these other signatures.** If systems utilizing these other signatures were to be deployed, methods that might be used in an attempt to hide or mask opacity and gamma-ray signatures would not necessarily defeat these systems under development—complicating any terrorist attempts to smuggle nuclear weapons or SNM into the United States. At the same time, these future systems tend to be more costly and complex than current systems; whether the added benefits are worth the added costs is a political decision.

Detection systems have their limits. Systems to detect SNM at close range, such as at ports and land border crossings, are generally not applicable to detection of terrorists smuggling a weapon across a remote stretch of border. But that is not a flaw of the detection system. Detectors can work at "points," i.e., places where people or cargo may enter the United States legally. There, detectors attempt to find SNM or weapons that may be hidden in cargo. In contrast, at "lines," the vast distances between "points" along coasts or borders, any entry is illegal, so interdiction is a matter of law enforcement. Effective intrusion detection systems (TV cameras, seismic monitors) coupled with a CONOPS that provides rapid response may suffice, though they have a long way to go to become effective. At the same time, standoff radiation detection systems that have yet to be developed, mounted along borders at natural choke points or smuggling routes, might be of value for this mission.

Observations on Technical Progress and Congress

Congress has supported a broad portfolio of detection R&D projects that has created a pipeline with technologies expected to become available for operational systems from near-term to long-term. These systems exploit many signatures in addition to those of currently-deployed systems, offering Congress the prospect of improved detection capability and a broader menu of choices. Several technical factors may influence the choice among technologies to support. For example: (1) Projects will advance at different rates. (2) Projects may benefit differently from an advance in a related technology. (3) As a project moves from research to development to deployment, cost and capability may vary from early projections.

Congress may wish to reevaluate current deployment decisions if it concludes that significantly more capable systems will be available in a year or two. Of course, any such decision would depend on comparing such factors as cost, footprint, ease of use, production rate, and the like for competing systems, and caution is necessary in assessing contending claims.

On the other hand, **it is difficult for Congress to choose among contending technologies.** Each requires evaluation in such terms as cost, scan time, ease of use, reliability, schedule, footprint, radiation exposure, spatial resolution, and ability to thwart shielding. Yet these data are difficult to obtain. Some are proprietary. Some are unknown: schedules may slip and costs rise, or technical advances may cause the opposite to occur. Developers of a technology tend to be its advocates, and see the strengths of their technology and a path to overcome its weaknesses. Even if these data can be obtained, it is necessary to weight data elements to support a choice among contending technologies. With many variables to be traded off against each other, how are weights to be assigned, and who decides? And can this weighting system function despite weaknesses in the data?

Congress has focused much attention on preventing terrorists from smuggling nuclear weapons or SNM into the United States in cargo containers. For example, P.L. 110-53, Implementing Recommendations of the 9/11 Commission Act of 2007, Section 1701, states, "A container that was loaded on a vessel in a foreign port shall not enter the United States (either directly or via a foreign port) unless the container was scanned by nonintrusive imaging equipment and radiation detection equipment at a foreign port before it was loaded on a vessel." While terrorists might attempt to smuggle in a nuclear weapon by other means, **developing technology to scan containers at seaports is a reasonable place to start.** Container-scanning technology can be modified for use in other situations, such as monitoring air cargo containers or passenger cars, which are easier to scan because they can contain much less shielding. Developing and deploying detection equipment for use at seaports ensures ruggedness and ease of use adequate for real- world applications, and forces governments at all levels to plan CONOPS.

More generally, some could argue that it is impossible to prevent terrorists from smuggling nuclear weapons into the United States, so there is no point in spending large sums in a futile effort. **Congress has rejected that approach, and has appropriated, in total, billions of dollars to deploy available systems and to support R&D on advanced technologies. Supporters of the R&D and deployment approach assert that it offers several advantages.**

- It has provided some capability quickly, increasing the odds of detecting weapons or SNM. An important example of this is the rapid deployment of passive radiation detectors to scan maritime cargo containers.
- This limited detection capability would help deter terrorists and would complicate plans to smuggle in weapons or SNM.
- Initial deployments provide data of use to subsequent deployments. They help refine what throughput, robustness, etc., front-line inspectors require of a system. They help

refine CONOPS. They help define desirable features of an architecture. These results can make future technologies, systems, and architectures more effective.

- It has created an R&D pipeline that is intended to generate a steady stream of new technologies and systems.
- The resulting improvements in individual technologies, operations, and architectures can improve overall system effectiveness.
- As technologies become more capable, they can plug gaps in the current architecture. For example, remote detection might offer a way to monitor choke points in the United States or overseas that terrorists might pass through in transporting SNM or weapons.

Congress may wish to address gaps and synergisms in this portfolio. For example:

Gaps: Several systems may use helium-3 tubes for neutron detection, yet the supply is limited. Alternatives are available, but the longer developers take to switch to these alternatives, the longer it would take to deploy their systems because of the need to incorporate different detectors, modify algorithms, and test the revised system. Other gaps include sensors that can detect SNM at long range (e.g., over 100 m), sensors that can operate autonomously in remote areas, and large but inexpensive detectors that can distinguish SNM from other radioactive material.

Synergisms: A component, algorithm, or material developed for one system may be applicable to another. Projects are under way to develop more sensitive materials to detect gamma rays and neutrons. These materials can be used in systems that induce fission in SNM. Their improved sensitivity permits a smaller source (e.g., an accelerator) to generate the interrogation beam, reducing cost, complexity, and radiation dose. Similarly, if detector material offers only fair resolution of gamma ray spectra, then peaks in a spectrum may blur, requiring a complicated algorithm to deal with the uncertainties. Sharper resolution from improved materials would reduce these uncertainties, permitting simpler algorithms to be used. More powerful computers could support faster, more powerful algorithms, reducing scan time, false positives, and false negatives.

Minimizing gaps and maximizing synergisms have the potential to lead to more capable systems faster and at lower cost. Companies that considered using helium-3 for neutron detection might expedite deployment and reduce costs by sharing effort to develop an alternative neutron detector for their common use. Information on progress in developing more sensitive detector material would permit companies to incorporate such materials into their systems sooner, also speeding deployment and lowering costs. Is there a way that development could be shared or licensed so that companies, especially those working on government-funded projects, could avoid duplicating effort? And could this be done while retaining the benefits of competition?

In considering the Advanced Spectroscopic Portal, Congress and the Government Accountability Office examined in detail whether DNDO had followed proper procedures for testing competing systems. **An alternative means by which Congress could address testing is to direct the executive agency in charge of a system to conduct specified tests.** These tests would need to be designed, and perhaps overseen, by experts not affiliated with the relevant agency, company, or laboratory. Congress has ample access to the technical expertise required. The relevant congressional committees could consult with individual experts or with groups that have a long history of providing independent technical advice to the government, such as the American Association for the Advancement of Science, the JASON defense advisory group, the National Academy of Sciences, the National Council on Radiation Protection and Measurements, and the National Institute of Standards and Technology. In this way, Congress could seek a fair comparison between systems on variables of interest, such as scan times or the ability to detect specified targets in containers with specified cargoes, enhancing confidence in the test results and decisions based on them. Other alternatives exist. Congress could require DHS to establish an independent test and evaluation unit; obtain an outside review of DHS test and evaluation procedures; require DNDO to provide detailed reporting of each step in the acquisition process as it occurs; or provide for an external review of a program.

Observations on Technical Progress and Terrorism

Ongoing improvement in U.S. detection capabilities produces uncertainties for terrorists that seem likely to increase over time, adding another layer of deterrence beyond that of the capabilities themselves. Capability of fielded equipment may be upgraded. Terrorists may not know the capability or availability of future detectors. More advanced technologies should improve detection capability. It should be harder for terrorists to evade new systems than current ones. Detection may affect terrorists in another way. A nuclear weapon would be of immense value to them. Therefore, increasing the risk of detection would have a much greater deterrent effect for them than for drug smugglers, where detection and confiscation of drugs are part of the cost of doing business. The multiplication of technical obstacles to a successful terrorist attack may thus help deter an attack or an attempt to undertake a project to launch one.

At the same time, it is important to **avoid the "fallacy of the last move."** Herbert York, a former Director of Defense Research and Engineering, coined this term to argue that in the Cold War nuclear arms race, one side's actions typically led to countervailing actions by the other side.[120] The same principle applies to nuclear detection. This report suggests that some U.S. detection systems nearing readiness for deployment are more capable than current detectors. Yet if terrorists were to attempt to bring a nuclear weapon or SNM into the United States, they could use various techniques to evade detection by such systems, though these techniques might complicate the plot and increase the risk of detection by non-technical means. Further, the threat might increase in various ways, such as if new terrorist groups emerged or if more nations built nuclear power plants or nuclear weapons. For such reasons, Congress has funded, and executive agencies are pursuing, R&D with short- and longer-term time horizons. Also for such reasons, the global nuclear detection architecture may need to be updated from time to time. Thus, while the United States has an immense technical advantage in a competition of detection vs. evasion, and a pipeline of increasingly more capable technologies, it is important to recognize not only the dynamic aspects of advances in detection capabilities but also the dynamic aspects of the competition.

APPENDIX. THE PHYSICS OF NUCLEAR DETECTION[121]

What Is to Be Detected?

Detectors must detect complete weapons, which can be quite small. During the Cold War, the United States made 155 mm and 8 inch (diameter) nuclear artillery shells. The United States made even smaller atomic demolition munitions, and there have been reports of Soviet-era "suitcase bombs." A weapon that terrorists fabricated without state assistance would surely be less sophisticated and, as a result, probably much larger. Detectors must also detect the types of uranium and plutonium used in nuclear weapons. The type of uranium used in weapons is harder to detect than plutonium because it emits much less radiation; it is also much easier to fabricate into a weapon component. It is important to detect small quantities of these materials in order to interdict stolen and smuggled materials because small quantities suffice to fuel a bomb. According to a widely-quoted report by five nuclear weapon scientists from Los Alamos National Laboratory, it would take 26 kg of uranium, or 5 kg of plutonium (both of types discussed later) to fuel an atomic bomb.[122] These masses would fit into cubes 11.2 cm or 6.3 cm, respectively, on a side. The ability to detect even smaller masses would help thwart nuclear smuggling. How is it possible to find weapons or materials among the vast amount of cargo that reaches the United States each day? Fortunately, there are many clues.

Background

Photons

Nuclear detection makes extensive use of photons. Photons are packets of energy with no rest mass and no electrical charge. Electromagnetic radiation consists of photons, and may be measured as wavelength, frequency, or energy; for consistency, this report uses only energy, expressed in units of electron volts (eV).[123] Levels of energy commonly used in nuclear detection are thousands or millions of electron volts, keV and MeV, respectively. The electromagnetic spectrum ranges from radio waves (some of which have photon energies of 10^{-9} eV), through visible light (a few eV), to higher-energy x-rays (10 keV and up) and gamma rays (mostly 100 keV and up). An x-ray photon and a gamma-ray photon of the same energy are identical.

Gamma rays originate in processes in an atom's nucleus. Each chemical element has two or more isotopes. Isotopes of an element have the same number of electrons, and thus in most cases similar chemical properties, but different numbers of neutrons in their nuclei, and thus different nuclear properties. Each radioactive isotope emits gamma rays in a unique spectrum, a plot of energy levels (horizontal axis) and number of gamma rays detected at each energy level (vertical axis). These spectra are a series of spikes at particular energy levels.[124] **Figure 1** and **Figure 2** show the spectra of uranium-235 and plutonium-239, respectively. Such spectra are the only way to identify an isotope outside a well-equipped laboratory. A detector with a form of "identify" or "spectrum" in its name, such as Advanced Spectroscopic Portal or radioactive isotope identification device, identifies isotopes by their gamma-ray spectra.

X-rays originate in interactions with an atom's electrons. Many detection systems use x-ray beams, which can have higher energies than gamma rays and thus are more penetrating. X-ray beams are often generated through the *bremsstrahlung* process, German for "braking radiation," which works as follows. An accelerator creates a magnetic field that accelerates charged particles, such as electrons, which slam into a target of heavy metal. When they slow or change direction as a result of interactions with atoms, they release energy as x-rays whose energy levels are distributed from near zero to the energy of the electron beam. They do not exhibit spectral peaks like gamma rays. This difference is important for detection.

Radioactivity

Radioactive atoms are unstable. They decay by emitting radiation, principally alpha particles (a helium nucleus consisting of two neutrons and two protons, thus having a double positive charge), beta particles (electrons or positrons, the latter being electrons with a positive charge), and gamma rays (high-energy photons). These forms of radiation are of differing relevance for detection. Alpha particles, being massive (on a subatomic scale) and electrically charged, are easily stopped, such as by a sheet of paper or an inch or two of air. Beta particles, while much lighter and faster, are also electrically charged and are stopped by a thin layer of material.[125] Gamma rays have no charge and can penetrate much more material than can alpha or beta particles. Depending on their energy, they may travel through several hundred feet of air. When an atom decays by emitting an alpha particle or beta particle, it transforms itself into a different element; it does not do so when it emits a gamma ray. Gamma-ray emission typically follows alpha or beta decay. As discussed in more detail below, each radioactive isotope that emits gamma rays does so in a spectrum of energies unique to that isotope. For example, the spectrum of U-235 has a prominent peak at 186 keV.

In addition to these typical means of radioactive decay, atoms of some heavy elements fission, or split into two smaller atoms. Of the naturally-occurring isotopes, only U-238 spontaneously fissions with an appreciable rate (about 7 fissions per second per kg). One by-product of fission is the emission of neutrons (typically 2-3 neutrons per fission). Neutrons have no electrical charge and can penetrate dense materials, as well as many tens of meters of air.

Fissile material

Some isotopes of heavy elements fission spontaneously or when struck by neutrons or high-energy photons, emitting neutrons and gamma rays in the process. U-235 and Pu-239 are unique in that neutrons of any energy can cause them to fission; they are called "fissile." Neutrons of much higher energies are required to cause other isotopes to fission. This characteristic of U-235 and Pu-239 allows them to support a nuclear chain reaction. Fissile material is essential for nuclear weapons; U-235 and Pu-239 are the standard fissile materials used in modern nuclear weapons. The Atomic Energy Act of 1954 designates them as "special nuclear material" (SNM).[126]

Plutonium is not found in nature. It is produced from uranium fuel rods in a nuclear reactor and is separated from uranium and other elements using chemical processes. Weapons-grade plutonium (WGPu) is at least 93% Pu-239. In contrast, uranium in nature consists of 99.3% U-238 and 0.7% U-235, with very small amounts of other isotopes. Enriching it in the isotope 235 for use as nuclear reactor fuel or in nuclear weapons cannot be done through chemical means because isotopes of an element are nearly chemically identical,[127] so other means must be used. For example, uranium may be converted to the gas uranium hexafluoride and placed in centrifuges specially designed to separate U-235 from U-238 based on the very slight differences in the weight of individual molecules. Uranium enriched to 20% in the isotope 235 is termed highly enriched uranium, or HEU; for use in nuclear weapons, uranium is typically enriched to 90% or so, though lower enrichments could be used. For purposes of this report, "HEU" is used to refer to uranium of 90% enrichment. HEU may also be produced from material that has been in a nuclear reactor. HEU produced in this manner contains small amounts of another isotope, U-232, which, as we shall see, is easier to detect than is U-235.

Detection

Nuclear detection uses neutrons and high-energy photons in various ways. Because they can penetrate different materials, they are the main forms of radiation by which most radioactive material can be detected passively, by "listening" for signals coming out of a container without sending signals in. Because of their penetrating properties, they can be used in an active mode to probe a container for dense material. X-rays or gamma rays are used for radiography, that is, creating an opacity map like a medical x-ray. Neutrons of any energy level, and photons above 6 million electron volts (MeV), can be shot into a container to induce fission in SNM. Fission results in the emission of neutrons and gamma rays, which can be detected. Gamma rays can also be used to identify a radioactive source. Neutrons, in contrast, do not have a characteristic energy spectrum by which an isotope can be identified, and it is difficult to measure their energy, though the presence of neutrons in certain situations, as discussed below, can indicate that SNM is present.

Another characteristic of radioactive materials important for detection is the rate at which a material decays. The half-life of an isotope, or the time it takes for half the atoms in a sample to decay, is an indicator of the rate of decay, with shorter half-lives indicating faster decay. The half-lives of cobalt-60, plutonium-239, and uranium-235 are 5.3 years, 24,000 years, and 700 million years, respectively.[128] [129] Even if a source emits high-energy gamma rays, it will be difficult to detect if it emits only a few of them. Thus type, energy level, and quantity of radiation are important for detection.

Shielding and background radiation

Different materials attenuate neutrons and gamma rays in different ways. Heavy, dense materials like lead, tungsten, uranium, and plutonium have a high atomic number (the number of protons in the nucleus), or "Z." High-Z materials attenuate gamma rays efficiently.[130] In contrast, neutrons are stopped most efficiently by collisions with the nuclei of light atoms, with hydrogen being the most effective because it has about the same weight as neutrons.[131] The element with the nucleus closest in weight to a neutron is hydrogen, which in its most common isotope consists of one proton and one electron. Thus hydrogen-containing material like water, wood, plastic, or food are particularly efficient at stopping neutrons; other low-Z material is less efficient at stopping neutrons, but nonetheless more effective than high-Z material. Conversely, gamma rays are less attenuated by low-Z material and neutrons are less attenuated by high-Z material.

Different amounts of material are needed to attenuate gamma rays depending on their energy level. Gamma rays from WGPu are sufficiently energetic and plentiful that it is more difficult to shield WGPu than HEU. In contrast, as explained in the footnote, an inch of lead would render gamma rays from U-235 essentially undetectable, though as discussed later other uranium isotopes that may be present in HEU are more readily detectable.[132] Indeed, 186-keV gamma rays from U-235 have so little energy that many are absorbed by the uranium itself, a process known as self-shielding, so that the number of gamma rays emitted by a piece of U-235 depends on surface area, not mass.

Unclassified demonstrations performed at Los Alamos National Laboratory for the author in June 2006 indicate how shielding and self-shielding impair the detection of low-energy gamma rays from HEU. The demonstrations used a top-of-the-line detector that had an excellent ability to identify materials by their gamma-ray spectra.[133] In the first demonstration, the detector picked up gamma rays from a thin sheet of HEU foil at perhaps 30 feet away and quickly identified them as coming from HEU. The foil had a large surface area and little thickness, so there was little self- shielding. In the second, the detector gradually picked up gamma rays from a marble of HEU as it was brought closer to the detector. Because the marble had much more thickness and much less surface area than the HEU foil, there was considerable self-shielding, greatly reducing the gamma-ray output. In the third, the marble of HEU was placed in a capsule of a high-Z material, lead, perhaps 1 cm thick, and the detector picked up nothing even when the capsule was touching the window of the detector.

Sources of radiation other than SNM complicate detection. Background radiation from naturally-occurring radioactive material, such as thorium, uranium, and their decay products such as radon, is present everywhere, albeit often in trace amounts. Cosmic rays generate low levels of neutrons. Some legitimate commercial goods contain radioactive material, such as ceramics (which may contain uranium), kitty litter (which may contain thorium and uranium), and gas mantles made of thorium oxide. Other radioactive isotopes are widely used in medicine and industry. Finally, a terrorist group might conceivably place radioactive material in a shipment containing a weapon or SNM chosen so as to mask the unique gamma-ray spectrum of SNM by presenting a spectrum of several known innocuous materials with peaks to interfere with those of SNM or that have an intensity much higher than SNM.

Signatures of Plutonium, Highly Enriched Uranium, and Nuclear Weapons

For purposes of this report, a signature is a property by which a substance (in particular, SNM) may be detected or identified. A nuclear weapon or its fissile material may be detected by various signatures, some of which are discussed next.

Atomic number and density

Atomic number, abbreviated "Z," is the number of protons in an atom's nucleus. For example, the Z's of beryllium, iron, and uranium are 4, 26, and 92, respectively. Z is a property of individual atoms. In contrast, density is a bulk property, and is expressed as mass per unit volume, e.g., grams per cubic centimeter. The densities of beryllium, iron, and uranium are 1.848, 7.874, and 19.050 g/cc, respectively. At its most basic, density measures how many neutrons and protons (which constitute almost all of an atom's mass) of a substance are packed into a volume. In general, the densest materials are those of high Z. These properties may be used to detect uranium and plutonium. Uranium is the densest and highest-Z element found in nature (other than in trace quantities); plutonium has a slightly higher Z (94), and its density varies from slightly more to slightly less than uranium, depending on its crystal structure. Some detection methods discussed in Chapter 2, such as effective Z, make use of Z; and some, such as radiography and muon tomography, make use of Z and density combined.

Opacity to photons

An object's opacity to a photon beam depends on its Z and density, the amount of material in the path of the beam, and the energy of the photons. Gamma rays and x-rays can penetrate more matter than can lower-energy photons, but dense, high-Z material absorbs or scatters them. Thus a way to detect an object, such as a bomb, in a container is to beam in x-rays or gamma rays to create a radiograph (an opacity map) like a medical x-ray.

Presence of gamma rays beyond background levels

Background gamma radiation is ubiquitous. Since many materials, including SNM, emit gamma radiation, elevated levels of gamma radiation may or may not indicate the presence of SNM.

Presence of neutrons beyond background levels

Cosmic rays and naturally-occurring uranium generate a very low background flux of neutrons. Most materials do not emit neutrons spontaneously, but HEU and plutonium do. The spontaneous emission rates for 1 kg of plutonium and 1 kg of of HEU are 60,000 neutrons per second and 3 neutrons per second, respectively.[134] As a result, neutrons above the cosmic ray background coming from a cargo container would be suspicious.[135] For HEU, however, the rate is not too different from the background and thus is not a strong signature.

Gamma ray spectra

Each isotope has a unique gamma ray spectrum. For example, uranium-235 produces gamma ray peaks at several dozen discrete energy levels. This spectrum of energies is well characterized for each isotope, and is the only way to identify a particular isotope outside a well-equipped laboratory. As a result, any detector with a variant of "spectrum" or "identify" in its name, such as Advanced Spectroscopic Portal or radioactive isotope identification device, relies on identifying isotopes by their gamma-ray spectra.

HEU presents other gamma ray signatures as well. HEU contains some U-238, which produces a gamma-ray peak at an energy of 1.001 MeV. While these gamma rays are energetic, they would be hard to detect unless the detector is very close to the uranium because they are emitted at a very low rate, and could easily be missed because trace amounts of naturally-occurring uranium, such as in clay and soil, also generate 1.001 MeV gamma rays. HEU derived from spent nuclear reactor fuel rods also contains small amounts of uranium-232, which is formed when uranium is bombarded with neutrons in a nuclear reactor. Uranium-232 decays through a long decay chain of short-lived isotopes to thallium-208, which has a gamma ray of 2.614 MeV, one of the highest-energy gamma rays produced by radioactive decay, so it is distinctive as well as highly penetrating; it takes 2.041 cm of lead to attenuate half the gamma rays of that energy. Thallium-208 is also a decay product of naturally-occurring thorium-232. U-232 decays very much faster than U-235 or U-238 (half-lives of 69 years, 700 million years, and 4.5 billion years, respectively), and thallium-208 decays even faster (half-life of 3 minutes), so even a very small amount of U-232 produces many gamma rays.[136] Similarly, WGPu presents various gamma ray signatures because it is a mix of several isotopes of plutonium and their decay products.

Time pattern of neutrons and gamma rays

SNM is unique in that it can fission when struck by low-energy ("thermal") neutrons. Like some other materials, it also fissions when struck by high-energy gamma rays. In a sufficiently large mass of SNM, the neutrons (usually two or three) released by the fission of one atom cause other atoms to fission, releasing more neutrons in a chain reaction.[137] SNM also fissions spontaneously, and neutrons released by these fissions have a non-negligible probability of causing other SNM atoms to fission. Characteristic products of fission offer indications that SNM is present. These products include neutrons that may be emitted over periods ranging from nanoseconds to many seconds, whether as a result of spontaneous fission or of fission induced by gamma rays or neutrons, and gamma rays emitted within nanoseconds of induced fission.

Prompt gamma rays and neutrons

When U-235 and Pu-239 fission, they release a nearly instantaneous burst of 2 or 3 neutrons and 6 to 10 gamma rays. These prompt neutrons are emitted in a continuum of energies, with an average of about 1 to 2 MeV, and are termed fast or high-energy neutrons. The prompt gamma rays are also emitted in a spectrum of many narrow lines. Only SNM will fission when struck by low-energy neutrons, so a beam of low-energy neutrons that results in a burst of neutrons and gamma rays indicates the presence of SNM. A beam of high-energy gamma rays (with energy greater than 6 MeV) will also cause SNM to fission. However, that beam will also cause other materials to fission, including natural uranium, so emission of a burst of neutrons and gamma rays resulting from interrogation by a high-energy gamma ray beam is a possible, but not a definitive, indicator by itself of the presence of SNM.

Delayed gamma rays and neutrons

When U-235 or Pu-239 atoms fission, they split into two smaller fission fragments in any of approximately 40 ways for each isotope, resulting in "[s]omething like 80 different fission fragments" for U-235 or Pu-239.[138] These fission fragments are unstable and decay radioactively into isotopes of various elements. Fission is a statistical process, so that fissioning of a great many U-235 or Pu-239 atoms produces a complex mixture of some 300 isotopes of 36 elements.[139] These isotopes have a great range of half-lives, from a small fraction of a second to millions of years, but the isotopes with a half-life greater than approximately 30 years emit only very low levels of radiation. This process produces thousands of times more gamma rays than neutrons. Since much cargo consists of low-Z material and since gamma rays penetrate low-Z cargo much more readily than do neutrons, many more gamma rays than neutrons resulting from fission of SNM escape containers holding such cargo. Higher-Z cargo will attenuate the gamma rays more than the neutrons.

Some of the gamma rays have energies exceeding those of thallium-208, 2.614 MeV, the highest energy typically observed in natural backgrounds. "Their high energy makes this gamma radiation a characteristic of fission, very distinct from normal radioactive background that typically produces no gamma radiation exceeding an energy of 2.6 MeV."[140] Note that some other isotopes, such as U-238 and Pu-240, are "fissionable," that is, they can undergo fission only when struck by high-energy (fast) neutrons. The high-energy gamma rays resulting from fission are a strong indicator of the presence of SNM.[141] The intensity of the neutron and gamma-ray flux over a short period, caused by rapid decay of many of the fission products, and the prompt response to a probe, are distinctive signatures as well.

There is another time-delay signature. A neutron beam makes atoms of some other elements radioactive, in particular transforming some atoms of stable oxygen-16 to radioactive nitrogen-16. Researchers at Lawrence Livermore National Laboratory conducted experiments in which they bombarded a target of natural uranium (99.3% U-238, 0.7% U-235) inside a cargo container with a neutron beam, and recorded the gamma ray spectrum resulting from radioactive decay. After they turned off the neutron beam, they found that the high-energy portion of the spectrum was dominated by gamma rays from the decay of nitrogen-16 for the first 15 seconds, and after that the dominant signal was from the decay of radioactive fission products, with an average half- life of about 55 seconds.[142] This time difference is an indicator of the presence of SNM.

Differential die-away

Interrogation of SNM with a beam of neutrons or high-energy photons to induce fission produces another unique signature. While the beam may cause neutrons to be emitted immediately through various nuclear reactions (e.g., fission), materials other than SNM will not support a nuclear chain reaction. In contrast, even a subcritical mass of SNM can sustain a chain reaction for a short time.

As a result, fission in a multi-kilogram block of SNM will continue to produce neutrons for a short time after the beam has been turned off, with the intensity and duration of the neutron flux depending on the amount of SNM and the cargo loading. This delayed neutron time signature is called differential die-away, is measured on the order of a thousandth of a second after the beam is turned off, and is specific to U-235 and Pu-239 (and, rarely, other fissile isotopes). This technique depends on the detection of the prompt fission signal, but hydrogenous materials such as those found in cargo tend to attenuate this signal, and there may be background neutrons, so that some difficult scans may require more time, possibly two minutes, and some may not be feasible.

Fission chain time signature

A subcritical mass of SNM is too small to support a supercritical chain reaction because too many neutrons escape the SNM for the number of neutrons to increase exponentially. Nonetheless, chain reactions do occur in SNM, triggered by a neutron from spontaneous fission or a background neutron. These chain reactions may last several to dozens of generations, producing a burst of neutrons and gamma rays over some billionths of a second. No other material produces this signature. In contrast, most background neutrons and gamma rays arrive at a detector in a random pattern. The one exception is that neutrons generated as cosmic rays strike matter also tend to be generated in bursts; work is under way to try to differentiate between bursts of neutrons induced by cosmic rays and those generated by fission chains. Detection of this signature is therefore a strong sign of the presence of SNM. Unlike differential die-away or delayed neutrons and gamma rays, this signature can be detected with passive means provided the SNM is not well shielded. This technique places great demands on detector technology but can be done with state- of-the-art electronics.

Chapter 2 discusses in detail two other signatures—deflection of muons and nuclear resonance fluorescence and absorption—and their detection.

Detecting Signatures of a Nuclear Weapon or SNM

Overview: How are signatures gathered, processed, and used?

Detection involves using detector elements to gain data, converting data to usable information through algorithms, and acting on that information through concept of operations, or CONOPS. Detectors, algorithms, and CONOPS are the eyes and ears, brains, and hands of nuclear detection: effective detection requires all three.

Since photons or neutrons have no electrical charge, their energy is converted to electrical pulses that can be measured. This is the task of detectors, discussed next. The pulses are fed to algorithms. An algorithm, such as a computer program, is a finite set of logical steps for solving a problem. For nuclear detection, an algorithm must process data into usable information fast enough to be of use to an operator. It receives data from a detector's hardware, such as pulses representing the time and energy of each photon arriving at the detector. It converts the pulses to a format that a user can understand, such as displaying a gamma ray spectrum, identifying the material creating the spectrum, or flashing "alarm." Every detector uses one or more algorithms. Improvements to algorithms can contribute as much as hardware improvements to detector capability. Algorithms may be improved in many ways, such as by a better understanding of the physics of a problem, or by improving the computers in detection equipment so they can process more elaborate algorithms.

CONOPS may be divided into two parts. One specifies how a detection unit is to be operated to gain data. How many containers must the unit scan per hour? How close would a detector be to a container? Shall the unit screen cargo in a single pass, or shall it be used for primary screening, with suspicious cargo sent for a more detailed secondary screening? A second part details how the data are to be used. What happens if the equipment detects a possible threat? Which alarms are to be resolved on-site and which are to be referred to off-site experts? Under what circumstances would a port or border crossing be closed? More generally, how is the flow of data managed, in both directions?[143] What types of intelligence information do CBP agents receive, and how do data from detection systems flow to federal, state, and local officials for analysis or action? While this report does not focus on CONOPS because is not a technology, it is an essential part of nuclear detection.

How detectors work

A discussion of how detectors work is essential to understanding the capabilities and limits of current detectors and how detectors may be improved. Detecting each signature of a nuclear weapon or SNM requires a detector that is appropriate for that signature. Further, there is a hierarchy of gamma ray detectors. The simplest can only detect the presence of gamma radiation. The next step up, detectors with low energy resolution, have a modest capability to identify an isotope by its gamma ray spectrum. Next, detectors with high energy resolution have very accurate isotope identification capabilities. More sophisticated detector systems can also identify the presence of SNM by the time pattern of gamma rays released when such material fissions. The most sophisticated detector systems can produce an image showing where each gamma ray came from.

Detectors require a signal-to-noise ratio sufficient to permit detection. That is, they must be able to extract the true signal (such as a gamma-ray spectrum) from noise (spurious signals caused, for example, by background radioactive material or by imperfect detectors or data-processing algorithms). Two concepts are central to gamma-ray detector sensitivity: detection efficiency and spectral resolution. The former refers to the amount of signal a detector records. One aspect of detection efficiency is the fraction of the total emitted radiation that the detector receives. Radiation diminishes according to an inverse square law; that is, the intensity of radiation (e.g., number of photons per unit of area) from a source is inversely proportional to the square of the distance from the source.[144] Since a lump of SNM emits radiation in all directions, moving a detector closer to SNM, or increasing its size, increases efficiency. Reducing the cost of the active material in a detector may increase efficiency by making a larger sensor area affordable. Another aspect is the fraction of the radiation striking the detector that creates a detectable signal. For example, a detector that can absorb 90% of the energy of photons striking it is more efficient than one that can absorb 10%. A more efficient detector will collect information faster, reducing the time it takes to screen a cargo container.

Spectral resolution refers to the sharpness with which a detector presents energy peaks in a radiation spectrum. A graph of the gamma-ray spectrum of a radioactive isotope plots energy levels along the horizontal axis of the graph and the number of counts per unit time at each energy level along the vertical axis. A perfect device would record the energy levels of a gamma- ray spectrum as a graph with vertical "needles" of zero width because each radioactive isotope releases gamma rays only at specific energy levels. In practice, however, detectors are not perfect, and 186-keV gamma rays will be recorded as a bell curve centered on 186 keV. The narrower the spread of the bell curve,[145] the more useful the information is. Polyvinyl toluene (PVT), a plastic that is widely used in radiation detectors because it can be fielded in large sheets at low cost, is sensitive but has poor resolution, i.e., extremely wide bell curves for each gamma-ray energy level. As a result, while PVT can detect radiation, the peaks from gamma rays of different energy levels blur together, making it impossible to identify an isotope. **Figure 3** makes this point; it shows the spectra of 90% U-235 and background radiation as recorded by a PVT detector. At the other extreme, high-purity germanium (HPGe) produces very sharp peaks, permitting clear identification of specific isotopes. These detectors are expensive, heavy, have a small detector area, and must be cooled to extremely low temperatures with liquid nitrogen or a mechanical system, making them less than ideal for use in the field. However, mechanically cooled HPGe detectors weighing some 2.5 kg are being developed for field use.[146] **Figure 4** shows the spectrum of Pu-239 as recorded by various types of detectors with better resolution than PVT.

Various means can improve detector sensitivity.[147] One type of semiconductor detector crystal is cadmium-zinc-telluride, or CZT. The peak on the far right of each spectrum[148] in **Figure 5** shows improvement in the resolution of the gamma-ray spectrum for cesium-137 (a radioactive isotope) taken with different CZT detectors that, for the years indicated, were at the high end of sensitivity. The top line shows a spectrum taken with a CAPture device developed by eV Products (1995-1998); the middle line shows a spectrum taken with a coplanar-grid device developed by Lawrence Berkeley National Laboratory (2000-2003); and the bottom line shows a spectrum taken with a 3-D device developed by the University of Michigan (2008). Better CZT crystals and better ways to overcome limitations of these crystals have both improved sensitivity in various ways:

- Researchers have been able to grow larger crystals. CZT crystal volume for the three devices was 1.00 cc in 1995-1998, 2.25 cc in 2000-2003, and 6.00 cc at present. Larger crystals are more efficient, i.e., they can capture more photons, and more of the energy of individual photons, permitting more counts per unit time (i.e., more data).
- Crystal quality has improved. A more uniform crystal structure and fewer impurities allow for better transport of the photon-induced electrical charge through the crystal and thus more accurate determination of the energy of each photon.
- University of Michigan researchers have constructed three-dimensional arrays of CZT crystals, permitting their detector to determine the 3-D coordinates of each individual gamma ray photon as it interacts with the CZT crystal, in turn permitting location as well as identification of gamma-ray sources.
- Electronics have improved. Researchers have made significant progress in reducing the noise inherent in electronic circuits (application-specific integrated circuits) that translate signals from the interaction of photons with CZT into a form in which algorithms can process them. Reducing the noise in these circuits permits more accurate measurement of gamma-ray energy. For example, a circuit developed in 2007 by Brookhaven National Laboratory has improved energy resolution substantially, and other advances in detector electronics in the last few years enable electronic components to compensate for defects in the crystals (analogous to adaptive optics in astronomy).
- Algorithms to reconstruct the signal from gamma rays have improved, also permitting more accurate measurement of gamma ray energy.

Another factor that affects the ability to detect SNM is the time available for a detector to scan a container or other object, often called "integration time." Detectors build up radiography or tomography images, or gamma-ray spectra, over time. More time enables a detector to have more photons per pixel (in the case of radiography) or per bin (in the case of gamma-ray spectra), or more muons per voxel. More time also enables a neutron detector to detect more neutrons and measure their times of arrival, as discussed below, helping to determine if the neutrons are generated by SNM or by background materials. More time thus provides better data, which provides for better separation of signal from noise, better separation of different sources of radiation, fewer false alarms, and a better chance of detecting and identifying shielded threat material. **Figure 14** illustrates how one system builds up an image over time. From a physics perspective, then, increasing integration time improves the accuracy of the result, but from a port operator's perspective, longer integration time impedes the flow of commerce, which costs money, so a balance must be struck between these two opposed goals. This balance may be stated formally in a concept of operations (discussed in more detail below), which specifies how, among other things, a detection system will be operated; detection equipment must be designed to operate within the time required, and port operations must allow that amount of time for scans.

Still another means of improving the ability to detect SNM is to increase the spatial resolution of a detector. According to DTRA,

> This is easily demonstrated in the example of a shielded versus unshielded radiation detector. Unshielded detectors are sensitive to radiation impinging on it in all directions, which is characteristic of the nature of naturally-occurring background radiation. By adding shielding, a detector's field-of-view can be controlled, and background radiation levels reduced, increasing the signal-to-noise ratio for the detector in the direction the detector is aimed.[149]

Detecting gamma rays

Gamma rays do not have an electrical charge, but an electrical signal is needed to measure them. There are two main ways by which a gamma ray can be turned into electrical energy. One is with a scintillator material, such as PVT or sodium iodide. When a single higher energy photon, such as a gamma ray, strikes the scintillator and interacts with it, the scintillator emits a large number of photons of lower energy, usually visible light ("optical photons"). A photomultiplier tube (PMT) converts the optical photons to electrons, then multiplies the electrons to generate a measurable pulse of electricity whose voltage is proportional to the number of optical photons, which is in turn proportional to the energy deposited by the gamma ray. An electronic device called a multi-channel analyzer sorts the pulse into a "bin" depending on its energy and increases the number of counts in that bin by one. A software package then draws a histogram with energy level on the horizontal axis and number of counts on the vertical axis. The histogram is the gamma ray spectrum for that isotope.

In contrast, a semiconductor material, such as HPGe, turns gamma rays directly into an electrical signal proportional to the gamma-ray energy deposited. A voltage is applied across the semiconductor material, with one side of the material being the positive electrode and the other being the negative electrode. When a gamma ray strikes the material, it knocks some electrons loose from the semiconductor's crystal lattice. The voltage sweeps these electrons to the positive electrode. Their motion produces an electric current whose voltage is proportional to the energy of each gamma ray. Each pulse of current is then sorted into a bin depending on its voltage and the spectrum is computed as described above.[150]

This approach, with either type of detector, is used to detect the various gamma-ray signatures described earlier. However, the requirements for detecting time signatures varies somewhat. Because prompt gamma rays are emitted so quickly, identifying them requires the ability to record time of arrival to several billionths of a second. Delayed gamma rays of interest are generated over a period of tens of seconds, so the ability to record precise time of arrival is less important. Detecting fission chain time signature requires a high-efficiency detector because long fission chains are relatively rare. Thus to detect SNM rapidly, the detector must have a high efficiency for detecting every fission chain. While the delayed emissions from fission chains are too weak to detect passively, fission chain time signature focuses on detecting the prompt emissions from fission, which are stronger. High efficiency is also important for neutron and gamma-ray interrogation, but the emphasis is less stringent because far more fissions are induced (i.e., the signal is stronger). Detecting nuclear resonance fluorescence requires high-resolution detectors in order to differentiate between the various materials being analyzed.

Detecting neutrons

Neutrons, like photons, do not have an electrical charge, but the two interact with matter differently. Photons interact chiefly with electrons, while neutrons interact with atomic nuclei. As a result, neutrons are counted by a different process. A common neutron detector is a tube of helium-3 gas connected to a high-voltage power supply, with positively- and negatively-charged plates or wires in the tube. In its rest state, current cannot pass through the helium because it acts as an insulator. When a low-energy neutron passes through the tube, it is absorbed by a helium-3 atom, producing a triton (1 proton and 2 neutrons) and a proton. These particles are highly energetic and lose their energy by knocking electrons off other helium-3 atoms. Positively-charged ions of helium-3 move to the negative plate, while electrons move to the positive plate. Since electric current is the movement of charged particles, these particles generate a tiny electric current that is measured and counted. Neutrons are emitted as a continuum of energies. While the mean energy of each neutron spectrum varies somewhat from one isotope to the next, neutron energy spectra do not have lines representing discrete energies as with gamma rays. Moreover, neutrons lose energy as they collide with low-Z material, further blurring their spectra. Thus neutron spectra are of little value for identifying isotopes. Instead, the total neutron count is an important means of identifying SNM because only SNM gives off neutrons spontaneously in significant numbers, though some neutron background is generated mainly when cosmic rays knock neutrons off atoms. Several other methods of detecting SNM by neutron emission, discussed above, rely on the time pattern in which a group of neutrons arrives.

Several systems detect neutrons with tubes filled with helium-3 (He-3), a standard method. DOE obtains He-3 as a byproduct from the decay of tritium used in nuclear warheads. With the decades-long decline in numbers of warheads and a hiatus in tritium production for many years, there is little new supply of He-3. DOE plans to supply customers with 10,000 liters of He-3 a year, with a starting bid price expected to be around $72 per liter, and states, "This appears short of what customers are requesting."[151] (Russia sells He-3 to U.S. companies, but quantities are proprietary and not available.) Deploying He-3 neutron detection systems in large numbers would require a considerable amount of He-3. Customs and Border Protection (CBP) states that "based on our RPM [radiation portal monitor] deployments CBP would need approximately 2500 [detector] units to cover sea and land borders."[152] (Data for number of units needed for air cargo are not available.) Given the shortage and cost of He-3, deployment of neutron detectors using large amounts of He-3, or large numbers of units requiring small amounts of He-3, does not appear feasible.

Alternative neutron detection systems are possible. They include tubes coated with boron-10 or lithium-6, tubes filled with boron-10 trifluoride (a toxic gas), nanocomposite scintillators, and "neutron straws," thin tubes being developed commercially under sponsorship of the Defense Threat Reduction Agency.[153] Substituting any of these technologies for He-3 in a system would necessitate re-engineering the system's neutron detectors, revising algorithms, conducting tests, perhaps modifying the resulting system for operational conditions, and so on. Those changes have the potential to add delay and affect system performance (for better or worse), though given the high cost of He-3 (about $2.7 million for 38,000 liters) they might well reduce cost.

Detecting absorption or scattering of high-energy photons

Photons of sufficiently high energy can penetrate solid objects. Denser, higher-Z material within a solid object absorbs photons of lower energy and scatters photons of higher energy. For cargo scanning, a fan-shaped planar beam of photons is sent through a cargo container as the container passes through the beam, and a detector array on the other side consisting of semiconductor or scintillator material records the opacity of each pixel to the photons. An algorithm then creates a two-dimensional opacity map of the contents of the container and displays it as an image on a computer screen.

Increasing the energy of photons allows them to penetrate more material. Radiography is used to search cargo containers for terrorist nuclear weapons, among other things.[154] A radiograph would reveal clearly a large dense object, such as a nuclear weapon encased in lead shielding. Two limitations of radiography are noteworthy. First, radiographs do not detect radiation and thus do not specifically detect SNM, just high-density, high-Z material. Second, if a terrorist bomb is placed in a shipment of dense or mixed objects, the image of the bomb might be hidden or a radiographic equipment operator might not notice it. It would be much harder to detect a small piece of SNM using radiography than to detect a bomb.

Evasion of Detection Technologies

In order to understand the capabilities of detection systems, it is important to know their weaknesses as well as their strengths. However, detailed discussions of means of evasion tend to become classified. Some references are made throughout this report, but some are withheld to keep the report unclassified. In general, an enemy could use various means in an effort to defeat these technologies. For example, high-Z material absorbs and deflects gamma rays, low-Z material deflects neutrons, radiography might miss a small piece of SNM (especially if mixed in with other dense material), and reducing the apparent density and Z of SNM by mixing it with a low-Z substance reduces the deflection of muons.

Further, enemy attempts to defeat one type of detection system may complicate plans or make a plot more vulnerable to detection by other means, as several examples illustrate. (1) The use of multiple detection systems that detect different phenomena are harder to defeat than those detecting one phenomenon only. Placing a lead shield around a bomb in order to attenuate gamma rays from plutonium would create a large, opaque image that would be quite obvious on a radiograph. It is for this reason that Congress mandated, "A container that was loaded on a vessel in a foreign port shall not enter the United States (either directly or via a foreign port) unless the container was scanned by nonintrusive imaging equipment and radiation detection equipment at a foreign port before it was loaded on a vessel." This restriction is to apply by July 1, 2012.[155] (2) An enemy could attempt salvage fuzing, which would detonate a weapon if the weapon sensed attempts to detect it, such as with photon beams, or if it was tampered with. However, salvage fuzing has various shortcomings. It could result in a weapon detonating by accident or if it is scanned overseas. It could detonate a weapon in a U.S. port, where it would do much less damage than in a city center. (3) Enemy attempts to smuggle HEU into the United States in order to avoid detection of a complete bomb would require fabricating the weapon inside this nation, which in turn could require such activities as smuggling other weapon components and purchase of specialized equipment, and could run the risk of accidents (such as with explosives), any of which could provide clues to law enforcement personnel. For these reasons, it is important to view technology development not only as advances in capabilities per se but also in the context of an offense-defense competition.

End Notes

[1] Comparison would require a detailed review of hundreds of technology projects to determine which are most worthy of further examination; creating metrics to compare the selected projects; and obtaining accurate data for use in the metrics. Each of these tasks would require the work of many experts over many months.

[2] For further information, see CRS Report RL34574, *The Global Nuclear Detection Architecture: Issues for Congress*, by Dana A. Shea.

[3] See Charles Ferguson and William Potter, *The Four Faces of Nuclear Terrorism*, Monterey, CA, Center for Nonproliferation Studies, 2004; Michael Levi, *On Nuclear Terrorism*, Cambridge, MA, Harvard University Press, 2007; and Carson Mark et al., "Can Terrorists Build Nuclear Weapons?," Nuclear Control Institute, Washington, DC (no date).

[4] Some materials can fission only when struck by fast neutrons; fissile materials are the only materials that can fission when struck by slow as well as fast neutrons.

[5] The Atomic Energy Act of 1954, 42 U.S.C. 2014, defines SNM as uranium enriched in the isotopes 233 or 235 or plutonium. The Nuclear Regulatory Commission has not declared any other material to be SNM even though the Act permits it to do so. U.S. Nuclear Regulatory Commission. "Special Nuclear Material."

[6] Mark et al., "Can Terrorists Build Nuclear Weapons?"

[7] Roger Byrd et al., "Nuclear Detection to Prevent or Defeat Clandestine Nuclear Attack," *IEEE Sensors Journal*, August 2005, p. 594.

[8] Ibid.

[9] For further analysis of this topic, see CRS Report RL34070, *Fusion Centers: Issues and Options for Congress*, by John Rollins.

[10] Personal communication, Defense Threat Reduction Agency, August 8, 2008.

[11] HEU that has been through a nuclear reactor picks up small quantities of U-232, which decays through intermediate steps to thallium-208, which has a gamma ray of 2.614 MeV, one of the highest-energy gamma rays produced by radioactive decay. As a result, it is distinctive as well as highly penetrating, facilitating detection.

[12] P.L. 110-53, Implementing Recommendations of the 9/11 Commission Act of 2007, Section 1701, 121 Stat. 489.

[13] The principal investigator for this project, Edward McKigney, Senior Project Leader, Safeguards Science and Technology Group, Nuclear Nonproliferation Division, Los Alamos National Laboratory, provided detailed information for this section, personal communications, April-August 2008. Others have commented as well to provide alternative perspectives.

[14] Specifically, the number of gamma rays depositing all their energy increases as Z to the 4.5^{th} power.

[15] These properties include probability of a full energy deposition interaction; efficiency of converting deposited energy to light, the signal that is measured; transparency to photons generated in this way so the detector volume responds uniformly; and physical robustness.

[16] For example, dense packing of nanocrystals may change the material's mechanical flow. That would make it harder to process the material by the lowest-cost method, extrusion, but it could still be cast and machined to shape. PVT scintillator is cast and can be machined, suggesting that that approach offers low cost while producing large sheets of material.

[17] For example, the project needs an understanding of the structural features of the materials being synthesized that an instrument called a resonant Raman spectrometer could provide. McKigney states that this instrument would speed up development of the process to synthesize the scintillator material by providing direct information about what material has been synthesized. Currently, the project infers such information from measurements that McKigney states are harder to interpret.

[18] The principal developer of this algorithm, Dean Mitchell, Distinguished Member of the Technical Staff, Contraband Detection Organization, Sandia National Laboratories, provided detailed information for this section, personal communications, April-August 2008. Others have commented as well to provide alternative perspectives.

[19] For a technical discussion of GADRAS, see Dean J. Mitchell, "Variance Estimation for Analysis of Radiation Measurements," Sandia Report SAND2008-2302, April 2008.

[20] Dean J. Mitchell, "Analysis of Chernobyl Fallout Measured with a RAMP Detector," SAND87-0743-UC-32, Sandia National Laboratories, 1987.

[21] Dean J. Mitchell, "Analysis of Low-Resolution Gamma-Ray Spectra by Using the Unscattered Flux Estimate to Search an Isotope Database," Systems Research Report, Sandia National Laboratories, 1997.

[22] For further information, see U.S. Department of Homeland Security, Customs and Border Protection, "Laboratories and Scientific Services." Available at http://www.cbp.gov/xp/cgov/trade

[23] Personal communication, August 5, 2008.

[24] A compiler is a computer program that translates an application (such as GADRAS) into instructions that a computer can process.

[25] As another example, because of incompatibilities between VB6 and current generations of Fortran, components of GADRAS that are written in Fortran must be compiled using Compaq Fortran, another discontinued and unsupported programming language.

[26] Personal communication, August 5, 2008.

[27] Richard Wheeler, Lead for Homeland Security Analysis, Global Security Directorate, Lawrence Livermore National Laboratory, provided detailed information for this section, personal communications, April-August 2008. Others have commented as well to provide alternative perspectives.

[28] For a helpful interactive tutorial on ROC curves, see Anaesthetist.com, "The Magnificent ROC," at http://www.anaesthetist.com/mnm/stats/roc/Findex.htm.

[29] Lawrence Livermore National Laboratory, Radiological and Nuclear Countermeasures Program, "Radiation Detection Modeling and Operational Analysis: Benchmarks, Nuisance Source, Algorithm and Resolution Studies," LLNL-TR-401531, December 31, 2007, by Padmini Sokkappa et al., pp. 21-22 This source is a report marked by Lawrence Livermore National Laboratory, the originating organization, "For Official Use Only." That laboratory has approved CRS use of the information cited by this footnote.

[30] This paragraph is based on ibid., pp. i-ii.

[31] Radiological and Nuclear Countermeasures Program, "Radiation Detection Modeling and Operational Analysis," p. 17. This source is a report marked by Lawrence Livermore National Laboratory, the originating organization, "For Official Use Only." That laboratory has approved CRS use of the information cited by this footnote.

[32] Ibid., pp. 20-21.

[33] Joel Rynes, Program Manager, CAARS Program, Domestic Nuclear Detection Office, Department of Homeland Security, provided detailed information for this section, personal communications, April-August 2008. Others, including L-3 Communications Corporation, have commented as well to provide alternative perspectives.

[34] "Prior to 9/11, not a single radiation portal monitor [RPM] and only 64 large-scale non-intrusive inspection [i.e., radiography] systems were deployed to our nation's borders. By October of 2002, CBP had deployed the first RPM at the Ambassador Bridge in Detroit." Testimony of Thomas Winkowski, Assistant Commissioner, Office of Field Operations, U.S. Customs and Border Protection, before the Senate Homeland Security and Governmental Affairs Committee, September 25, 2008. CBP has 1,145 RPMs and 209 large-scale radiography systems deployed as of October 23, 2008. Personal communication, Patrick Simmons, Director, Non-Intrusive Inspection, Customs and Border Protection, October 23, 2008. Radiography has been used for industrial applications (e.g., inspecting metal parts for cracks and voids) for decades.

[35] For information on these systems, see Rapiscan Systems, "Rapiscan Eagle P6000," at http://www.rapiscan.com/eagle-P6000.html, and Science Applications International Corporation (SAIC), "Safety & Security: VACIS Cargo, Vehicle, and Contraband Inspection Systems," at http://www.saic.com/products/security

[36] Jonathan Katz et al. argue that it would be more effective to have the beam interrogate a cargo container vertically rather than horizontally. "In innocent cargo long slender dense objects are packed with their longest axes horizontal, and dense cargoes are spread on the floor of the container. Therefore, near-vertical irradiation will only rarely show regions of intense absorption [of photons] in innocent cargo. In contrast, horizontal irradiation would often find this 'false positive' result, requiring manual unloading and inspection. Another advantage of downward near-vertical illumination is that the Earth is an effective beam-stop, combined with a thin lead ground plane, its albedo [the fraction of the beam that is reflected] is negligible and additional shielding would not be required." J.I. Katz et al., "X- Radiography of Cargo Containers," *Science and Global Security*, Vol, 5, No. 1, January 2007, pp. 49-56.

[37] Katz states, "The real issue in radiographing thick targets is not the source strength, but scattering in the target. If you don't have a narrow Bucky collimator matched to the geometry, the signal (increase of attenuation of unscattered photons showing the high-Z object ...) is swamped by forward-scattered photons from the mass of solid material. The small feature disappears from the image (rather like exposing undeveloped camera film to the light)." Personal communication, August 8, 2008. For cargo screening using radiography, a "Bucky collimator" is a piece of high-Z metal (such as tungsten) placed just in front of a photon detector element. The metal has a small hole precisely aligned with the direction of the interrogation beam in order to eliminate most photons that have been scattered by the cargo and that could obliterate small features of the image, such as a piece of SNM. The resulting increase in signal-to-noise ratio greatly improves the detector's ability to pick out suspicious cargo.

[38] The scattering mechanisms are much more complicated than can be described here. For details, see Glenn Knoll, *Radiation Detection and Measurement*, third edition (New York, John Wiley & Sons, 2000), pp. 48-53, 308-3 12.

[39] The threshold value of Z>72 is used because elements between Z=57 (lanthanum) and Z=72 (hafnium), inclusive, are very rare in commerce, making 72 a reasonable boundary between high Z and lower Z elements for the purposes of CAARS.

[40] For a periodic table of the elements, see "WebElements" at http://www.webelements.com/.

[41] For a technical discussion of dual-energy radiography, see S. Ogorodnikov and V. Petrunin, "Processing of Interlaced Images in 4B10 MeV Dual Energy Customs System for Material Recognition," Physical Review Special Topics—Accelerators and Beams, Volume 5, 104701 (2002).

[42] DNDO dropped its original goal of scanning 120 containers per hour. Testimony of Vayl Oxford, Director, Domestic Nuclear Detection Office, before the Senate Homeland Security and Governmental Affairs Committee, September 25, 2008.

[43] U.S. Department of Homeland Security. Domestic Nuclear Detection Office. "Cargo Advanced Automated Radiography (CAARS) Program Update Brief," presented to the Senate Committee on Homeland Security and Government [sic] Affairs, February 25, 2008, by William Hagan, Assistant Director, DNDO, Transformational and Applied Research, slide 4. This source is a briefing marked by the Domestic Nuclear Detection Office, the originating organization, "For Official Use Only." That office has approved CRS use of the information cited by this footnote.

[44] "Opening Statement of Mr. Vayl S. Oxford, Director, Domestic Nuclear Detection Office, Department of Homeland Security, before the Senate Homeland Security and Governmental Affairs Committee," September 25, 2008, p. 4.

[45] Information provided by Ira Reese, Executive Director, Laboratories and Scientific Services, and Patrick Simmons, Director, Non-Intrusive Interrogation, both of Customs and Border Protection, personal communication, October 23, 2008.

[46] Information provided by Joel Rynes, personal communication, October 28, 2008.

[47] Domestic Nuclear Detection Office. "Cargo Advanced Automated Radiography (CAARS) Program Update Brief," slides 3, 18. This source is a briefing marked by the Domestic Nuclear Detection Office, the originating organization, "For Official Use Only." That office has approved CRS use of the information cited by this footnote.

[48] Ira Reese, Executive Director, Laboratories and Scientific Services, and Patrick Simmons, Director, Non-Intrusive Interrogation, both of Customs and Border Protection, provided information in this paragraph, personal communication, October 23, 2008.

[49] Testimony of Thomas Winkowski, Assistant Commissioner, Office of Field Operations, U.S. Customs and Border Protection, before the Senate Homeland Security and Governmental Affairs Committee, September 25, 2008.

[50] Information provided by Joel Rynes, personal communication, October 28, 2008.

[51] Personal communication, July 11, 2008.

[52] U.S. Department of Homeland Security. "Advanced Technology Demonstration for Shielded Nuclear Alarm Resolution." Broad Agency Announcement 08-102 for the Domestic Nuclear Detection Office, Transformational and Applied Research Directorate, March 2008, p. 22, available at https://www.fbo.gov/files/6f6/6f6f82cc03fe6fcbf298a7e3903a15b7.doc?i=640d9818cdf76fa884b8c064b8f193 76.

[53] Rex Richardson, Vice President and Principal Scientist, Science Applications International Corporation, provided detailed information for this section, personal communications, June-August 2008. Others have commented as well to provide alternative perspectives.

[54] See http://www.accuray.com/.

[55] Electron beams of the levels discussed here generate *bremsstrahlung* photons of a wide spectrum of energies. Most of the photons are of relatively low energy, well below 1 MeV. They contribute almost nothing to the radiographic image but can represent a major portion of the radiation exposure to personnel. To remove or "filter" them, a piece of copper is placed in front of the photon beam. The material is thick enough to stop lower-energy photons but not higher- energy ones.

[56] Fission also releases prompt gamma rays, but they cannot be counted by a system of this configuration because of interference by the large number of beam photons.

[57] Personal communication, August 3, 2008. See also J.I. Katz, "Detection of Neutron Sources in Cargo Containers," *Science and Global Security* 14, 2006: 145-149.

[58] Jeffrey Illig, Program Manager, AS&E CAARS Program, and Stephen Korbly, Director of Science, Passport Systems, provided detailed information for this section, October 2008. Others have commented as well to provide alternative perspectives.

[59] The EZ-3D unit can operate without a radiography system. However, the AS&E CAARS system would use an EZ-3D system and a radiography system together.

[60] A Rhodotron is a circular electron accelerator manufactured by Ion Beam Applications. Unlike most linear accelerators, it generates electron beams in continuous waves rather than in pulses.

[61] Sodium iodide detectors are used instead of detectors with a higher resolution, such as high-purity germanium. Since the EZ-3D system requires only that the detectors count photons, as opposed to identifying isotopes by their gamma-ray spectra, detectors with medium energy resolution suffice, and are much less costly than germanium detectors. Since the Rhodotron generates an electron beam in a continuous wave, and thus generates an x-ray beam in a continuous wave with no large spikes in the number of x-rays, the sodium iodide detector is able to record all the time. In contrast, a beam that turned on and off several thousand times a second would produce large spikes in the number of photons, overloading the detector.

[62] Personal communication, October 27, 2008.

[63] Personal communication, October 8, 2008.

[64] Personal communication, October 22, 2008.

[65] Information provided by Ion Beam Applications, October 24, 2008. These figures are for a Rhodotron for use in x- ray mode, i.e., with a target for the electron beam to strike to create *bremsstrahlung* x-rays. As of October 30, 2008, 1 = $1.2923.

[66] Personal communication, October 22, 2008. The home page for the Bates Linear Accelerator Center is http://mitbates.lns.mit.edu/bates/control

[67] Michael Sossong, Director of Nuclear Research, Decision Sciences Corporation, and Guest Scientist, Los Alamos National Laboratory, provided detailed information for this section, personal communications, April-July 2008. Others have commented as well to provide alternative perspectives.

[68] A lump of plutonium, whether shielded or not, seems an implausible threat because it would be very difficult for terrorists, by themselves, to fabricate a bomb using plutonium. It would be even harder for them to fabricate such a bomb inside the United States using plutonium they had smuggled in because they would need to take added measures to avoid detection.

[69] Personal communication from Michael Sossong, Los Alamos National Laboratory, April 23, 2008.

[70] Jonathan Katz, Karol Lang, and Roy Schwitters, "Muon Tomography—The Future of Vehicle and Cargo Inspection," report prepared for Decision Sciences Corporation, July 19, 2007, p. 5.

[71] Ibid.

[72] Ibid., p. 4.

[73] Ibid., pp. 9-10.

[74] A voxel is a volume element, analogous to a two-dimensional pixel, or picture element.

[75] Deflection is influenced both by Z and density. A muon is more likely to interact with a larger atom (higher Z) than with a smaller one. A muon is also more likely to interact with atoms the closer they are packed together (density). Scattering density combines Z and density into one unit.

[76] Katz et al., "Muon Tomography—The Future of Vehicle and Cargo Inspection," pp. 19-20.

[77] Information provided by Simon Bedford, General Manager, Guardian MT, Decision Sciences Corporation, personal communication, October 27, 2008.

[78] Personal communication, July 15, 2008.

[79] Personal communication, Mell Stephenson, Executive Director of Government Programs, Decision Sciences Corporation, June 16, 2008.

[80] Decision Sciences Corporation, "Decision Sciences Corporation Announces Agreement with Los Alamos National Laboratory to Collaborate on Homeland Security," press release, May 3, 2007, p. 1.

[81] Information provided by Simon Bedford, General Manager, Guardian MT, Decision Sciences Corporation, personal communication, October 27 and November 4, 2008.

[82] U.S. Department of Energy. Office of Declassification. "Drawing Back the Curtain of Secrecy: Restricted Data Declassification Policy, 1946 to the Present, RDD-1." June 1, 1994. Item V (C) (2) (s), at https://www.osti.gov/ opennet/forms.jsp?formurl=document/rdd- 1 /drwcrtf3.html#ZZ1.

[83] The Royal Society, "Detecting Nuclear and Radiological Materials," RS policy document 07/08, March 2008, p. 6, at http://royalsociety.org/displaypagedoc.asp?id

[84] Personal communication, July 31, 2008.

[85] Personal communication, July 15, 2008.

[86] For further information, see State University of New York at Buffalo, Department of Nuclear Medicine, Center For Positron Emission Tomography, "Positron Emission Tomography," at http://www.nucmed.buffalo petdef.htm.

[87] Personal communication, Michael Sossong, July 12, 2008.

[88] See "Feds Fight Threat of Small-Boat Terror Strikes," CNN, April 27, 2008, at http://www.cnn.com/2008/US/ 04/27/ small.boat.terror.ap/index.html#cnnSTCText.

[89] Stephen Korbly, Director of Science, Passport Systems, Dennis McNabb, Deputy Division Leader, N Division, Lawrence Livermore National Laboratory, and Glen Warren, Senior Research Scientist, Pacific Northwest National Laboratory, provided detailed information for this section, April-August 2008. Others have commented as well to provide alternative perspectives.

[90] For an image of minerals fluorescing under ultraviolet light, see Glenbow Museum, Calgary, Alberta, "Fluorescent Minerals," at http://www.glenbow.org/collections/museum/minerals/flourescent.cfm.

[91] For a description of this process, see Passport Systems, Inc., "Technology," at http://www.passportsystems.com/tech.htm.

[92] Personal communication, Leon Feinstein, DNDO, August 8, 2008. This technology would accelerate particles to high energies over much shorter distances than are possible at present. For example, existing accelerators can increase the energy of particles (e.g., electrons) by 5 to 10 MeV per meter; the new technology might increase that to over 150 MeV per meter, making for a much more compact accelerator.

[93] Collimation filters many forms of electromagnetic radiation so that only photons traveling in a certain direction are allowed through. In the case of x-rays or gamma rays, a collimator is typically a plate of lead or tungsten with many small parallel holes drilled through it.

[94] William Bertozzi and Robert Ledoux, "Nuclear Resonance Fluorescence Imaging in Non-Intrusive Cargo Inspection," *Nuclear Instruments and Methods*, B241, 820 (2005), p. 7.

[95] Personal communication, Stephen Korbly, Passport Systems, June 9, 2008.

[96] Personal communication, Stephen Korbly, Passport Systems, August 8, 2008.

[97] In order to identify the composition of a voxel with high confidence, the germanium detector must receive enough photons. If the material is dense, a longer scan time may be needed to accumulate the required number of photons.

[98] Personal communication, Stephen Korbly, Passport Systems, Inc., August 23, 2008.

[99] Personal communication, August 9, 2008.

[100] Personal communications, August 9 and 21, 2008.

[101] David Jordan and Glen Warren, "Simulation of Nuclear Resonance Fluorescence in Geant4," in Institute of Electrical and Electronics Engineers, Nuclear Science Symposium Conference Record, 2007, vol. 2, pp. 1185-1190.

[102] This model uses the Monte Carlo N-Particle Transport Code, developed by Los Alamos National Laboratory; see http://mcnp-green.lanl.gov/.

[103] The Rhodotron is made by Ion Beam Applications, a Belgian company. The company states, "The Rhodotron® is a recirculating accelerator where electrons gain energy by crossing a coaxial-shaped accelerating cavity several times. This original design makes it possible to operate the machine in continuous mode for maximum efficiency and throughput." http://www.iba.be/industrial

[104] Information provided by Ion Beam Applications, October 24, 2008.

[105] Dr. G. Peter Nanos, Jr., Associate Director for Research and Development, Defense Threat Reduction Agency (DTRA), Major Brad Beatty, USAF, Branch Chief, Standoff Detection Branch, Nuclear Detection Technology Division, DTRA, and Dr. Luc Murphy, Research Scientist, Locate and ID Branch, Nuclear Detection Technology Division, DTRA, and others at DTRA provided detailed information for this section, personal communications, April- July 2008. Others have commented as well to provide alternative perspectives.

[106] Personal communication, August 5, 2008.

[107] The Royal Society, "Detecting Nuclear and Radiological Materials," RS policy document 07/08, March 2008, p. 5, available at http://royalsociety.org/displaypagedoc.asp?id

[108] Personal communication, August 5, 2008.

[109] "Photonuclear" refers to a nuclear reaction caused by a photon, in this case fission of SNM induced by high-energy photons.

[110] Data cannot be gathered when the beam is on, for there would be far more photons from the beam than photons from fission of SNM, making it difficult or impossible for the detectors to identify the latter.

[111] Jonathan Katz states: "In principle, a sufficiently long and narrow collimator could produce an arbitrarily narrow x- ray beam. The price paid is that there is very little energy in the beam. For example, if the initial divergence is 30 degrees (plausible for an x-ray beam), then a beam collimated to 10 cm in diameter at 200 m (5 X 10^-4 radian, or 2.5 X 10^-7 steradian) will contain about 2.5 X 10^-7 of the source's energy and power. The rest is absorbed in the collimator or scattered into a diffuse flux; those photons are of no use for detection. The result would be a tiny amount of energy on the target. Any signal from fission of SNM generated by that energy would be emitted roughly equally in all directions, so a 100 cm^2 detector collocated with an x-ray source 200 m away would only pick up about 2 X 10^-8 of the signal. If that same detector were 5 m from the target but the accelerator were 200 m from the target, the detector would pick up only 3 X 10^-5 of the signal. Of course, background radiation isn't reduced at all, and would be several orders of magnitude stronger than the signal from fission. These estimates also ignored attenuation in the air. For a 10- MeV photon (a typical product of a 30-MeV accelerator), the beam is reduced by a factor of about 3 every 250 meters due to attenuation in the air, and the attenuation of 1-MeV fission gammas is about 5 times as great (a factor of 3 every 50 m). The attenuation of fission neutrons is about a factor of 3 every 150 m." Personal communications, September 29-October 8, 2008.

[112] The surface area of a sphere is $4\pi r^2$, where r is the radius. The surface area of a sphere with a 1-meter radius is 12.6 square meters. The surface area of a sphere with a 1,000-meter radius is 12.6 million square meters. Thus

a detector one meter square would receive one-millionth as many photons at a distance of 1 km as compared to a distance of 1 m.

113 DTRA states, "It is recognized that perhaps the single greatest challenge in successful implementation of this technology is the placement of high sensitivity (and specificity) detectors at ranges that may have to be significantly closer than the interrogation to be effective. These studies, to include modeling and experimentation are proceeding as part of the research program." Personal communication, August 5, 2008.

114 Personal communication, August 5, 2008.

115 The Royal Society, "Detecting Nuclear and Radiological Materials," RS policy document 07/08, March 2008, p. 1, available at http://royalsociety.org/displaypagedoc.asp?id

116 Personal communication, August 9, 2008.

117 Personal communication, July 19, 2008.

118 Personal communication, August 4 and 5, 2008; emphasis added.

119 Personal communication, August 1 and 8, 2008.

120 Herbert York, *Race to Oblivion: A Participant's View of the Arms Race*, New York, Simon and Schuster, 1970, p. 211.

121 John Valentine, Lawrence Livermore National Laboratory, provided invaluable assistance in explaining the science presented in this section, April-July 2008. Others reviewed and commented on this section as well.

122 Mark et al., "Can Terrorists Build Nuclear Weapons?"

123 An electron volt is a unit of energy used for measuring atomic and nuclear processes. One electron volt (eV) is equal to the amount of energy gained by a single unbound electron (one not part of an atom) when it accelerates through an electrostatic potential difference of one volt. It is equal to 1.6×10^{-19} Joules. For comparison, the energy release in the fission of one uranium atom is 200 million electron-volts, and the energy required to remove an electron from a hydrogen atom is 13.6 eV. Information provided by Defense Threat Reduction Agency, personal communication, August 5, 2008.

124 For the gamma ray spectra of various isotopes, see "Gamma-Ray Spectra of Isotopes," within the Radiochemistry Society website at http://www.radiochemistry.org/periodictable/gamma_spectra/. For the percentage distribution of the dozens of gamma rays from uranium-235, see "TORI Data" in "WWW Table of Radioactive Isotopes," available at http://ie.lbl.gov/toi/nuclide.asp?iZA=920235.

125 A half-centimeter of air will stop a beta particle emitted by tritium, while 0.04 cm of water or 2 mm of aluminum will stop a beta particle emitted by iodine-131. Eckhardt, "Ionizing Radiation—It's Everywhere," pp. 18, 19.

126 Under the Atomic Energy Act of 1954, P.L. 83-703, 42 U.S.C. 2014, SNM is uranium enriched in the isotopes 233 or 235 or plutonium. The Nuclear Regulatory Commission has not declared any other material to be SNM even though the Act permits it to do so. U.S. Nuclear Regulatory Commission. "Special Nuclear Material." Available at http://www.nrc.gov/materials/sp-nucmaterials.html.

127 The degree to which chemical properties of isotopes are similar "depends on the element. For hydrogen/deuterium the chemical differences are substantial; you cannot survive on heavy water. For other light elements they are small but produce measurable effects (the whole field of paleoclimatology is based on this). For uranium they are infinitesimal." Personal communication, Jonathan Katz, August 7, 2008.

128 U.S. Department of Energy. Office of Environmental Management. Integrated Data Base Report—1996: U.S. Spent Nuclear Fuel and Radioactive Waste Inventories, Projections, and Characteristics, revision 13, December 1997; table B.1, "Characteristics of important radionuclides," http://web

129 A more precise indicator of decay is specific activity, the number of curies per gram of material, where 1 curie = 3.7 x 10^10 disintegrations per second. Plutonium-241, for example, has a specific activity of 102 curies/gram, and its rapid radioactive decay makes it so hot that pieces of it glow red. Plutonium-239 has a specific activity of .062 curies/gram, while the corresponding figure for uranium-235 is .000002.

130 "The attenuation of gamma rays depends on the energy of the gamma ray (generally more energetic gamma rays penetrate better, though there are some exceptions), the density of electrons (generally nearly proportional to the mass density or specific gravity) and how tightly the electrons are bound to the nuclei (much more strongly for high-Z elements). The last factor is the most important, and is why lead is used in shielding." Personal communication, Professor Jonathan Katz, Department of Physics, Washington University in St. Louis, August 7, 2008.

131 As an analogy, when one billiard ball strikes another squarely, the first transfers its energy to the second and stops, while the second moves with about the same speed and direction as the first. By contrast, if a billiard ball strikes a bowling ball squarely, the bowling ball will move forward slightly and the billiard ball will bounce back with nearly the same velocity with which it struck the bowling ball.

132 It takes .074 cm of lead to block half the gamma rays with an energy of 186 keV. Thus, 1 inch (2.54 cm) of lead has 2.54/.074 = 32.3 such thicknesses for 186-keV gamma rays, so (1/2) to the 32.3 power, or 1.9×10^{-10}, of these gamma rays will penetrate 1 inch of lead. One kg of HEU emits 4×10^{7} gamma rays per second at 186 keV, ignoring absorption of the gamma rays by the uranium. (Source: Roger Byrd et al., "Nuclear Detection to Prevent or Defeat Clandestine Nuclear Attack," IEEE Sensors Journal, August 2005, p. 594.) Accordingly, one 186-keV gamma ray photon could be expected to escape 1 kg of HEU surrounded by an inch of lead every 500 seconds or so. Absorption of gamma rays by uranium would reduce this number considerably. Increasing the

amount of U-235 in a bomb-usable shape would not affect this calculation much because most gamma rays would be absorbed by the uranium and the amount of lead would increase as the surface area of the uranium lump increased. Further, gamma rays radiate in all directions. Since most detectors do not surround the object to be inspected, such as a cargo container, it would capture only a part of these gamma rays, further reducing the probability of detection.

[133] The detector used high purity germanium and was cooled with liquid nitrogen.

[134] Roger Byrd et al., "Nuclear Detection to Prevent or Defeat Clandestine Nuclear Attack," *IEEE Sensors Journal*, August 2005, p. 594.

[135] "There is an important caveat in this statement, and that is cargo containers at sea. The so-called "ship-effect" that results from higher neutron levels aboard ships due to cosmic ray interactions with iron and other ship contents can result in spurious neutron readings from cargo containers at sea." Information provided by Defense Threat Reduction Agency, personal communication, August 5, 2008. Another caveat is that while innocent neutron sources other than background are rare, there are some, such as californium-252, which is produced in nuclear reactors and is used as a laboratory neutron source.

[136] For data on the properties of isotopes, see Lawrence Berkeley National Laboratory. Berkeley Laboratory Isotope Project. "Exploring the Table of Isotopes." Available at http://ie.lbl.gov/education/isotopes.htm.

[137] Not every neutron will cause further fissions. For example, some may escape the mass of SNM, and some may strike impurities. In a subcritical chain reaction, each neutron results in fissions that generate, on average, less than one additional neutron; this reaction dies out. In a critical chain reaction, each neutron results in fissions that, on average, produce one additional neutron. In a supercritical chain reaction, each neutron results in fissions that, on average, produce more than one additional neutron. Such a reaction may be controlled, as in a nuclear reactor, releasing energy over months or years, or uncontrolled, as in an atomic bomb, releasing vast amounts of energy in a fraction of a second.

[138] U.S. Department of Defense and Department of Energy. *The Effects of Nuclear Weapons*, Third Edition, compiled and edited by Samuel Glasstone and Philip Dolan, Washington, U.S. Govt. Print. Off., 1977, p. 633.

[139] Ibid.

[140] D.R. Slaughter et al., "The 'Nuclear Car Wash': A Scanner to Detect Illicit Special Nuclear Material in Cargo Containers," UCRL-JRNL-202106, January 30, 2004, p. 4.

[141] They are not a definitive indicator, however, because there could be other sources of fission, such as californium- 252, and cosmic rays could induce fission.

[142] D.R. Slaughter et al., "The 'Nuclear Car Wash': A Scanner to Detect Illicit Special Nuclear Material in Cargo Containers," UCRL-JRNL-202106, January 30, 2004, pp. 6-7.

[143] For further analysis of this topic, see CRS Report RL34070, *Fusion Centers: Issues and Options for Congress*, by John Rollins.

[144] See, for example, "Inverse Square Law, General," at http://hyperphysics.phy-astr.gsu.edu/Hbase/Forces/isq.html.

[145] The spread is measured as "full width at half maximum," that is, the width of the curve measured halfway from the top to the bottom of the curve.

[146] Personal communication, Defense Threat Reduction Agency, August 8, 2008.

[147] This paragraph was prepared with the assistance of Aleksey Bolotnikov, Physicist, Brookhaven National Laboratory, Professor Zhong He, Department of Nuclear Engineering and Radiological Sciences, University of Michigan, and Ralph James, Senior Physicist, Brookhaven National Laboratory, July 2008.

[148] The peak in the energy spectrum corresponds to the total energy of an incoming photon completely stopped inside the detector. The continuum on the left side of the peak is caused when an incoming photon scatters in the detector, depositing an unpredictable fraction of its total energy. For such events, the information about the photon's energy is lost. Such events contribute to the background that may affect detector performance.

[149] Personal communication, August 8, 2008.

[150] For a simple discussion of how semiconductors work, see Marshall Brain, ""How Semiconductors Work," at http://www.howstuffworks.com/diode.htm. For further detail, see Knoll, *Radiation Detection and Measurement*, Chapter 11, "Semiconductor Diode Detectors," pp. 353-403.

[151] Information provided by Isotope Program, U.S. Department of Energy, personal communication, June 30, 2008.

[152] Information provided by Customs and Border Protection, Department of Homeland Security, personal communication, July 25, 2008.

[153] See Proportional Technologies, Inc., "Neutron Straws," at http://www.proportionaltech.com/neutron.htm.

[154] See, for example, Katz, J. I., Blanpeid, G. S., Borozdin, K. N. and Morris, C., "X-Radiography of Cargo Containers," *Science and Global Security*, Vol. 15, 1: 49-56.

[155] P.L. 110-53, Implementing Recommendations of the 9/11 Commission Act of 2007, Section 1701, 121 Stat. 489.

In: Detection of Nuclear Weapons and Materials
Editor: Pietro Egidi
ISBN: 978-1-60741-512-1

Chapter 3

The Global Nuclear Detection Architecture: Issues for Congress

Dana A. Shea

Summary

The U.S. government has implemented a series of programs to protect the nation against terrorist nuclear attack. Some of these programs predate September 11, 2001, while others were established since then. Most programs are within the Nuclear Regulatory Commission; the Departments of Defense, Energy, and State; and agencies that became part of the Department of Homeland Security (DHS) upon its creation, and they are focused on detecting the illicit acquisition and shipment of nuclear and radiological materials and protecting and securing nuclear weapons. These disparate programs have historically been viewed as lacking coordination and centralized oversight.

In 2005, the Domestic Nuclear Detection Office (DNDO) was established within the Department of Homeland Security to centralize coordination of the federal response to an unconventional nuclear threat. The office was codified in 2006 through the passage of the SAFEPort Act (P.L. 109-347) and given specific statutory responsibilities to protect the United States against radiological and nuclear attack, including the responsibility to develop a "global nuclear detection architecture." Determining the range of existing federal efforts protecting against nuclear attack, coordinating the outcomes of these efforts, identifying overlaps and gaps between them, and integrating the results into a single architecture are likely to be evolving, ongoing tasks.

The global nuclear detection architecture is a multi-layered system of detection technologies, programs, and guidelines designed to enhance the nation's ability to detect and prevent a radiological or nuclear attack. Among its components are existing programs in nuclear detection operated by other federal agencies and new programs put into place by DNDO. The global nuclear detection architecture is developed by DNDO in coordination with other federal agencies implementing nuclear detection efforts and this coordination is essential to the success of the architecture.

This architecture is a complicated system of systems. Measuring the success of the architecture relative to its individual components and the effectiveness of additional investments are challenges. The DNDO is developing risk and cost methodologies to be applied to the architecture in order to understand and prioritize the various nuclear detection programs and activities in multiple federal agencies.

Congress, in its oversight capacity, has shown interest in the development and implementation of the global nuclear detection architecture and in the decision-making process attendant to investments in it. Other issues that may be foci of congressional attention include the balance between investment in near-term and long-term solutions for architecture gaps, the degree and efficacy of federal agency coordination, the mechanism for setting agency investment priorities in the architecture, and the efforts DNDO has undertaken to retain institutional knowledge regarding this sustained architecture effort.

INTRODUCTION

Detection of and protection against illicit acquisition and use of special nuclear material (SNM) is a longstanding concern of the U.S. government.[1] Since the development of nuclear weapons, federal agencies have been involved in securing U.S. nuclear assets against diversion, theft, and attack. Similarly, concerns that terrorists or non-state actors might acquire a nuclear weapon or the materials necessary to construct one have led to federal efforts to track, detect, and secure such materials both domestically and abroad.

Preventing a terrorist or non-state nuclear attack within the United States involves more than detection of the nuclear weapon. Detection of nuclear and radiological material is one part of a larger system of deterrence, counterproliferation, and response activities. Intelligence information regarding the intent and capability of terrorist and non-state groups and law enforcement activities disrupting the formation and action of these groups play key roles in preventing initial acquisition of nuclear and radiological materials. Subsequent to the detection of nuclear or radiological materials, successful interdiction of these materials is another crucial step. Nevertheless, this report addresses only the global nuclear detection architecture, not programs focusing on events prior or subsequent to the detection opportunity.

The federal government has implemented a series of programs focused on detecting the illicit shipment of nuclear and radiological materials and protectingand securing nuclear weapons and material. Following the events of September 11, 2001, these programs were augmented by new programs focusing on preventing radiological and nuclear terrorism within the United States. Some of these new and existing efforts had overlapping goals, but they generally used different approaches to improve the detection and security of nuclear materials. These programs largely reside within the Departments of Defense, Energy, and State; agencies that became part of the Department of Homeland Security (DHS) upon its creation in 2002; and the Nuclear Regulatory Commission. Many of these agencies have both national and international roles in nuclear defense, protecting domestic nuclear assets while aiding in securing or detecting the transport of foreign nuclear material.

Programs established by the Departments of Defense and Energy and the Nuclear Regulatory Commission have focused on the security of nuclear and radiological materials.

For example, the Department of Energy (DOE) International Nuclear Materials Protection and Cooperation program aids in securing foreign special nuclear material.[2] The Department of Defense (DOD), through the Defense Threat Reduction Agency (DTRA), has enhanced the security and safety of fissile material storage and transportation in the former Soviet Union while dismantling and destroying associated infrastructure.[3]

Other programs have focused on detection of nuclear and radiological materials in transit, in order to detect attempts to illicitly transport a nuclear weapon or special nuclear material across borders. The DOE Second Line of Defense (SLD) program aids in establishing capabilities to detect nuclear and radiological materials in foreign countries at ports of entry, border crossings, and other designated locations.[4] The Department of State Export Control and Related Border Security Assistance Program undertakes similar efforts to provide radiation detection capabilities at border crossings.[5] Other programs are designed to detect nuclear and radiological materials in transit towards theUnited States, through screeningeither at foreign ports or at the U.S. border. For example, U.S. Customs and Border Protection uses both handheld and portal-based radiation monitoring to detect nuclear and radiological materials entering the United States.

Once created, DHS expanded the deployment of radiation detectors, both portal monitors through the Radiation Portal Monitor (RPM) program and handheld and portable detectors through the U.S. Coast Guard and Customs and Border Protection. The DHS Science and Technology (S&T) Directorate began research and development activities to develop an improved radiation detection portal and an integrated plan and structure for the use of federal radiation detection equipment. Additionally, DHS developed several overarching initiatives, such as the Container Security Initiative and the Secure Freight Initiative, to increase the likelihood that nuclear and radiological material or a nuclear weapon would be detected, identified, and interdicted during shipping. These initiatives built on other federal efforts, such as the DOE Megaports Initiative, which deploys radiation detection equipment and aims to increase detection of nuclear materials at ports of departure rather than at ports of entry.

The early post-September 11, 2001, efforts of the federal government, taking place in several agencies and departments, were criticized by experts who perceived that these activities were uncoordinated and insufficient to protect the United States from nuclear terrorism.[6] The Defense Science Board, among others, recommended that the federal government make a greater, more organized effort to protect the United States against the nuclear terrorism threat.[7]

Domestic Nuclear Detection Office

The Domestic Nuclear Detection Office (DNDO) was established by President Bush on April 15, 2005.[8] Established to centralize coordination of the federal response to an unconventional nuclear threat, DNDO is located within the Department of Homeland Security. Its first budget was requested as part of the S&T Directorate, but it was subsequently established as an independent office whose Director reports directly to the Secretary. The office was given statutory authority in the SAFE Port Act (P.L. 109-347) and given specific responsibilities to protect the United States against radiological and nuclear attack. Among these responsibilities is to develop, with the approval of the Secretary and in coordination with the Attorney General, the Secretary of State, the Secretary of Defense, and the Secretary of Energy, an enhanced global nuclear detection architecture with implementation under which (A) the Office will be responsible for the implementation of the domestic portion of the global architecture; (B) the Secretary of Defense will retain responsibility for implementation of Department of Defense requirements within and outside the United States; and (C) the Secretary of State, the Secretary of Defense, and the Secretary of Energy will maintain their respective responsibilities for policy guidance and implementation of the portion of the global architecture outside the United States, which will be implemented consistent with applicable law and relevant international arrangements.[9]

The development and implementation of a global nuclear detection architecture is a challenging endeavor. Because federal efforts to protect against nuclear attack are spread among multiple agencies, determining the full range of existing efforts, coordinating the outcomes of these efforts, identifying any overlaps and gaps between them, and developing an architecture integrating current and future efforts are likely to be evolving, ongoing tasks.

What is the Global Nuclear Detection Architecture?

Although the SAFE Port Act requires that DNDO establish an "enhanced global nuclear detection architecture," it does not define this term. A variety of interpretations are possible. For example, an architecture could be a collection of federal and nonfederal programs, a grouping of sensors or other technologies designed to detect nuclear and radiological material, a mechanism for collecting and distributing information about nuclear and radiological material, a framework for investment and prioritization of detection assets, or various combinations of the above and more.

The DNDO describes the global nuclear detection architecture as comprising several key elements: "a multi-layered structure of radiological/nuclear (rad/nuc) detection systems, deployed both domestically and overseas; a well-defined and carefully coordinated network of interrelationships among them; and a set of systems engineering-based principles and guidelines governing the architecture's design and evolution over time."[10] In implementing this definition, DNDO solicited information about existing programs from agencies involved in nuclear detection. The DNDO then performed a "net assessment" of federal nuclear detection capabilities. This assessment determined that 72 programs contributed in whole or in part to the existing global nuclear detection architecture, with total funding of more than $2.2 billion in FY2006.[11]

This existing global nuclear detection architecture includes programsat DHS,[12] the Department of Defense (DOD),[13] the Department of Energy (DOE),[14] the Department of State (DOS), and other agencies. According to DHS, before the formation of DNDO these programs were "a disparate patchwork of systems, distributed and implemented in recent years across multiple departments, jurisdictions and locations without any degree of coordination."[15] Organized by DNDO into a global nuclear detection architecture framework, the combined system of systems relies heavily on its technological component, with the deployment of radiation detectors at points of entry, commercial ports, and other border crossings key to its effectiveness.

Although much focus has been given to technologies to detect nuclear or radiological material that have been developed or procured by DNDO, the global nuclear detection architecture encompasses more than just these sensors. Other elements include the use of sensor data to inform decision-makers, effective reaction to a detection event, and interdiction of the detected nuclear or radiological material. Accordingto the Government Accountability Office, "combating nuclear smuggling requires an integrated approach that includes equipment, proper training of border security personnel in the use of radiation detection equipment, and intelligence gathering on potential nuclear smuggling operations."[16] Other experts have concluded that the deployment of radiation detectors needs to be highly integrated with other federal efforts, prioritized on identified threats, configured for flexibility and efficiency, and part of a global approach including international institutions.[17]

The DNDO has attempted to align existing federal programs so that their capabilities can be compared and integrated into an organizing framework that can help identify gaps and duplication. This framework consists of three partially overlapping layers with nine sub-layers. See **Figure 1**.

Layer	Sublayer	Example
Exterior	Foreign Origin	Foreign sites with nuclear material that could be misused.
Exterior	Foreign Transit	Illicit trafficking of nuclear material within the exterior layer
Exterior, Border	Foreign Departure	Foreign seaport with cargo containers destined for the U.S.
Exterior, Border	Transit to U.S.	Ships transporting cargo from overseas to U.S.
Border	U.S. Border	Official U.S. ports of entry and between official land and sea ports of entry
Interior	U.S. Origin	Hospital with nuclear medicine equipment or industrial site
Interior	U.S. Regional	Areas surrounding origins of nuclear material in the U.S.
Interior	Target Vicinity	Areas surrounding potential targets of nuclear attack
Interior	Target	Potential locations of nuclear attack within the U.S.

Source: CRS adaptation of Domestic Nuclear Detection Office, May 17, 2007, as reproduced in Department of Homeland Security, Office of Inspector General, *DHS' Domestic Nuclear Detection Office Progress in Integrating Detection Capabilities and Response Protocols*, OIG-08-19, December 2007.

Figure 1. Layers of the Global Nuclear Detection Architecture

The layers are distinguished geographically: interior, border, and exterior. The overlap between the exterior and border layers may make analysis of priorities between and within the layers more difficult. The sublayers correspond mainly to conceptual steps in the transportation of a threat object to a target.

The global nuclear detection architecture has a broad, international scope, so implementing it is difficult. Multiple agency initiatives and programs must be relied on to achieve the architecture's goals, and its effectiveness is dependent on many factors outside of DNDO's direct authority and control.

By categorizing existing programs in this architecture, DNDO can analyze federal nuclear detection capabilities, identifying gaps and vulnerabilities through which a potential adversary might be able to avoid detection. These gaps may be filled by redirecting existing efforts, increasing existing efforts, deploying available technology, and implementing research and development programs that develop solutions to such gaps.

Layered Defense

A layered, defense-in-depth approach to a global nuclear detection architecture was recommended by the Defense Science Board when considering how to protect DOD assets against unconventional nuclear threats.[18] Successful application of a layered defense provides multiple opportunities to detect and interdict threats. According to DNDO, "It is recognized that no single layer of protection can ever be one hundred percent successful," and a layered defense strategy acknowledges this difficulty.[19] If one sublayer fails to detect a threat, the next may succeed.

This increase in the likelihood of detection occurs in two different ways. In one case, a threat may avoid the detector in an outer layer, but then encounter a detector in an inner layer. In this case, having more detection rings makes it more likely that a detector is encountered. An example of this approach could be the use of random truck screening at weigh stations on interstate highways. The DNDO has explicitly attempted to incorporate such redundancy into its global nuclear detection architecture, identifying numerous areas where detection capabilities might be integrated into existing operations

> Examples include, but are not limited to, fixed and mobile detection systems integrated into commercial vehicle inspection activities, detection enabled law enforcement, and screening conducted for special events. Capabilities that may require additional operational costs include mobile teams sweeping of areas of concern, chokepoint screening at bridges and tunnels, roadway monitoring concepts, and options for reducing the risk posed by the small maritime craft pathway.[20]

Alternatively, a threat may encounter a detector in an outer layer that fails to detect it, but then may encounter a different type of detector in an inner layer that is more successful. In this case, it is the use of different detection technologies or procedures that provides the increased likelihood of success. Examples might include the screening of manifest information for shipments entering the United States, followed by the use of radiation detection equipment; the use of both radiation detectors and non-intrusive imaging technologies; or the physical search of a vehicle triggered by suspicious behavior even though a radiation detector did not detect any emitted radioactivity. Prior experience has shown that nuclear smuggling detection occurs not only through the raising of alarms by radiation detection equipment at borders, but also by intelligence information and through police investigations.[21]

An additional advantage to a layered system is that its multiple detection and interdiction opportunities may increase the number of steps that a terrorist group must take to evade detection. Because of these additional steps, the group may be more likely to be detected by other means unrelated to the global nuclear detection architecture. For example, if it is necessary to disassemble a nuclear device to avoid detection, the reassembly of the device within the United States might be prevented by unrelated law enforcement activities. Even better, the increased complexity of evading detection might deter an attacker from even attempting a particular type of attack.

The ability to correlate information from different layers may also enhance the detection capability of the global nuclear detection architecture. Fusion of data from the different layers may reveal patterns or information not apparent in any single layer. It is the intent of the global nuclear detection architecture to integrate detection and notification systems at the federal,state, and local level,[22] but accomplishing that goal may take significant time and effort, as procedures, technology, and data formats may need to be harmonized to allow easy information exchange.

Methodology and Metrics for Evaluation

A significant advantage to establishing a global nuclear detection architecture is that it provides a framework for analysis of the overall effectiveness of federal nuclear detection efforts. Thus the performance of programs in each layer of the architecture can be measured and judged within the context of the overall structure rather than in isolation. In this way, effectiveness and efficiency can be maximized for the architecture overall rather than for each program individually.

Decision-makers attempting to analyze the architecture effectiveness and efficiency will likely require a methodology and establishing metrics, qualitative or quantitative, for each layer or sublayer. The DNDO uses the global nuclear detection architecture framework to identify gaps in existing detection capabilities.[23] Methodology to map existing and future capabilities onto an analytical construct that is sensitive to changes in the architecture would provide a more robust tool for prioritization and assessment. According to DNDO, application of such a risk- and cost-based assessment methodology to radiological and nuclear countermeasures would be relatively new, and DNDO planned to validate the employed methodology in 2008 on the basis of independent peer review.[24]

In 2004, the DHS S&T Directorate requested studies for the development of such an analytic framework and the identification of appropriate metrics.[25] This work was transferred to DNDO upon its creation in 2005.[26] Since then, additional studies of general aviation and maritime pathways have expanded the analytic basis for assessment of investments in the global nuclear detection architecture.[27] The degree to which existing programs can be related to these analytic frameworks likely determines their utility and applicability. A notional analytic framework — one in which some elements of the framework may not reflect the actual systems in place or someparameters are estimated or extrapolated rather than based on empirical data — may not be adequate for deciding which programs to invest in, alter, or otherwise optimize for maximum effect within the framework. On the other hand, a framework derived only from existing programs may overvalue the existing assets while undervaluing the potential contributions of new programs.

A further concern with respect to analytic methodology is its ability to reflect the effects of both large and small changes. The global nuclear detection architecture is a multi-billion dollar enterprise composed of dozens of programs. A methodology that encompasses all these programs but omits significant detail may not be sensitive enough to reflect the impact of incremental changes. For example, some experts have advocated deployment of radiation detection sensors at specific sites based on identified smuggling routes and at ports of entry where nuclear and radiological materials are not adequately secure.[28] Ideally, an analytic methodology would provide the means to compare that strategy to other strategies, such as an increase in border detectors.

A methodology that addresses all programs in sufficient detail may be cumbersome to use and may not reflect the value of intangible concepts, such as deterrence or misdirection.[29] An approach involving analysis of selected components of the global nuclear detection architecture, rather than the architecture as a whole, may make the analytical methodology more tractable. However, this may lead to inaccuracy when considering areas that fall between the individual component analyses or when considering the overall context of the architecture. Optimizing the efficiency and effectiveness of each individual component may not optimize the overall architecture. Even if it does, such an approach may not be cost effective.

The DNDO has stated that it takes a systems engineering approach to developing and refining the global nuclear detection architecture. Such an approach attempts to optimize the overall performance of the architecture rather than optimizing any particular program within it. Treating the global nuclear detection architecture as a "system of systems" may efficiently develop an effective architecture, but such treatment requires clear metrics around which the system is to be engineered.

The DNDO may find identifying the appropriate metrics for evaluation challenging. Outcome-oriented metrics, such as the number of false positives resolved,[30] the number of threats found, or the number of vehicles cleared, may not be suitable for judging the effectiveness of the architecture, though these metrics may help determine the efficiency or completeness of the planned architecture. On the other hand, more desirable measures, such as the degree of risk reduction, may require a more complete understanding of global risk than is currently available.

Metrics based on analysis of the existing global nuclear detection architecture may have similar difficulties. If the existing architecture has insufficient detection capability or coverage, incremental improvement to that architecture may lead to a new architecture that still has insufficient detection ability or coverage. Conversely, if the performance of the architecture is acceptable, incremental improvements may not yield substantial benefit when compared with the incremental cost.

Vulnerability or gap analysis could be used to prioritize and assess architecture effectiveness.[31] A challenge for this approach is the difficulty of determining an acceptable level of detection and geographic coverage.

The DNDO has identified gaps in the global nuclear detection architecture and is attempting to develop options and strategies for reducing these gaps.[32] A vulnerability- or gap-based approachrelies on implementing or developing solutions to these vulnerabilities or gaps, but the determination of a vulnerability may rely on an assessment of whether the detection capabilities of the existing system are sufficient. This assessment of a sufficient or acceptable detection capability, unlikely to be 100%, may be open to debate.[33]

Finally, the nature of the terrorist nuclear threat, potentially a changing threat based on an intelligent adversary, implies that any metrics and methodology developed to assess the global nuclear detection architecture's effectiveness will need to evolve as the threat does. When advances in technology, new intelligence information, and other factors are considered, the effectiveness of the global nuclear detection architecture may need to be judged on active testing or "red teaming" of the architecture.[34] The results of such active testing may be misleading if the testing does not conform to the threat for which the architecture is designed. For example, if the architecture is designed to detect only large amounts of a nuclear material, testing it with a small amount of nuclear material may highlight current limitations but not address the effectiveness of the architecture at the tasks for which it is designed.[35] Moreover, a robust architecture containing sublayers with varying detection success rates may still provide sufficient protection against a particularthreat, evenif asingle sublayer is insufficiently protective. In order to validate the results of "red teaming" exercises, DNDO plans to "conduct and quantify assessment results in various directions, including scenario-based 'bottom-up' assessments, capabilities-based 'top-down' assessments, and complex metrics development."[36]

Priority Setting

Gaps and vulnerabilities in the global nuclear detection architecture, depending on their nature, may be addressed now or in the future. In some cases, no solution to gaps and vulnerabilities is currently available, and a solution will need to be identified through research and development. The DNDO has stated that "there are still key, long-term challenges and vulnerabilities in our detection architecture that require long-range, higher risk research programs that will need to be evaluated in terms of risk reduction, direct and indirect costs, operational feasibility, and other relevant decision factors."[37] In other cases, the available near-term solution is an incremental improvement over existing approaches. In these cases, decisions must be made about whether to invest in a near-term, incomplete solution, accept the presence of a gap or vulnerability and invest in a long-term program to develop a more complete solution, or do both. Choosing between these options requires an understanding of the risk posed by the existing vulnerabilities, the benefits available through the near- and long-term options, and their relative costs.

Decision-makers are faced with difficult choices when setting priorities for implementing the global nuclear detection architecture. In the case of existing programs, incremental increases in the performance of a system may be challenged on the basis of their perceived costs and benefits.[38] In the case of new programs, questions may arise about whether the effort expended on a new program would have been better used elsewhere. Finally, given that improvement of the global nuclear detection architecture is a multi-year project, one must determine which portion of the architecture to focus on at any given time.

A likely benefit of casting federal efforts at nuclear detection into the framework of a global architecture is the ability to prioritize, in a quantitative or qualitative fashion, across programs.[39] Even absent a rigorous methodology to discriminate finely between the results of different investments, the global nuclear detection architecture may be able to provide a rank ordering of vulnerabilities and gaps, and thus a rank ordering of investment priorities.[40] Thus, it may provide an interagency tool to analyze current technology options and R&D investments relative to the federal government's detection needs.

The DNDO analysis methodology underpinning the global nuclear detection architecture continues to undergo revision and refinement:

> In order to maximize the effectiveness of the FY 2008 edition of the [global nuclear detection architecture], DNDO will leverage the independent observation of a full peer review to ensure that the requirements called forth in the [global nuclear detection architecture] continue to reduce the risk from nuclear terrorism. Accordingly, risk and economic impact methodology documents (carefully prepared and reviewed to protect sensitive/classified vulnerability information) will be produced and subjected to broad peer review.[41]

This review and these documents have not been made public. Congress may wish to determine whether the review addresses congressional concerns and whether the underlying architecture methodology is sufficiently robust. Alternatively, Congress may wish to direct DNDO to perform a review of the analysis methodology through an open process.[42]

Similarly, a global nuclear detection architecture may be able to highlight regions or modalities where investment or additional focus may provide a steeper or quicker reduction of vulnerability. For example, following the development of the baseline global nuclear detection architecture, DNDO decided to focus efforts on addressing vulnerabilities associated with aviation and maritime domains.[43]

Interagency Coordination

As well as developing the global nuclear detection architecture, DNDO is also responsible for coordinating the activities of other federal agencies whose programs make up the global nuclear detection architecture. For the architecture to be successful, substantial interagency coordination must occur on the operational and policy levels.

Congress recognized the need for DNDO to have access to specific talent resident in other agencies. The SAFE Port Act authorizes the DHS Secretary to "request thatthe Secretaryof Defense, the Secretary of Energy, the Secretary of State, the AttorneyGeneral, the Nuclear Regulatory Commission, and the directors of other Federal agencies, including elements of the Intelligence Community, provide for the reimbursable detail of personnel with relevant expertise to [DNDO]."[44] Under this authority and that of the Intergovernmental Personnel Act (IPA),[45] DNDO has established a significant interagency workforce, including personnel from DOD, DOE, the Federal Bureau of Investigation, the Department of State, and the Nuclear Regulatory Commission, as well as intra-agency personnel from the Science and Technology Directorate, U.S. Customs and Border Protection, the Transportation Security Administration, and the U.S. Coast Guard.[46]

The DNDO uses the detailees and IPAs as part of its coordinating function. By using these experts as conduits back to their agencies, DNDO is able to draw on the expertise and address the needs and concerns of these agencies. The DNDO also has established a more senior policy coordinating body, the Interagency Coordination Council, to address higher level policy issues and further coordinate activities between agencies, but the extent to which this body is able to implement and develop new policy for the participating agencies is not known. The Interagency Coordination Council was reportedly used to develop the deployment strategy for the global nuclear detection architecture and studies of maritime and aviation threats.[47]

The successful operation of the Interagency Coordination Council is critical for oversight and implementation of the global nuclear detection architecture, but procedural and organizational issues may pose barriers to its success. The Director of DNDO may not be equal in authority to the officials in other agencies with whom he is coordinating. Other officials may have more or less control of budgets, activities, and policies. Additionally, other agencies may perceive the global nuclear detection architecture as a DNDO document, rather than as a consensus coordination document. If so, other agencies may not quickly adopt the premises or analytical constructs expressed as part of the global nuclear detection architecture, preferring to continue to operate under individual agency priorities.[48]

The DNDO also has implemented an Advisory Council consisting of officials from other DHS components. The DNDO uses the Advisory Council to solicit the opinions of and resolve intra-agency issues within DHS.[49]

Beyond the interagency activities organized within DNDO, coordination of DNDO activities with other portions of the federal government occurs within the White House through the Domestic Nuclear Defense Policy Coordinating Committee. This joint policy coordination body was created jointlybythe Homeland Security Council and the National Security Council and provides a high-level forum for the generation of guidance and coordination among federal agencies with responsibilities for nuclear defense, detection, and interdiction.[50] Other interagency planning activities, such as coordination of long-term research and development, occur through subcommittees of the National Science and Technology Council.[51]

Congress has identified such coordination and cooperation as a key issue for DNDO, and in the Department of Homeland Security Appropriations Act, 2007, withheld $15 million from DNDO until a memorandum of understanding had been established with each federal entity and organization participating in DNDO activities.[52] The DNDO entered into these agreements throughout FY2007.[53]

ISSUES FOR CONGRESS

Multiple agencies implement the global nuclear detection architecture, even though its development is located within a single agency. Congress, in its oversight role, may be able to assess agency implementation of the global nuclear detection architecture across the federal government and thus identify weaknesses or inefficiencies that may occur. Additionally, Congress may be uniquely positioned to address policy challenges. Mechanisms for policy setting, the establishment of funding levels within the global nuclear detection architecture, implementation of development plans for the architecture and the identification of solutions to gaps and vulnerabilities, and the continued maintenance of the global nuclear detection architecture all are issues that Congress may choose to address.

Priority Setting and Appropriateness of Funding Levels Within the Global Nuclear Detection Architecture

The annual federal investment in the global nuclear detection architecture is spread across multiple agencies and across the layers and sublayers of the global nuclear detection architecture. The appropriate balance of funds in each of the different layers and sublayers, as well as between the different programs and agencies, is likely an issue of policy interest. When Congress established the Cooperative Threat Reduction program in 1991, it focused on securing nuclear materials at their source and preventing these materials from being transferred into non-state hands.[54] These continuing programs represent significant investment in the exterior layer of the architecture. More recently, DNDO and Congress have focused on the border layer of the global nuclear detection architecture. The DNDO has invested in Advanced Spectroscopic Portal (ASP) and Cargo Advanced Automated Radiography System (CAARS) technologies to improve the ability to detect nuclear and radiological materials at the borders,[55] and Congress has mandated the improved screening of cargo containers shipped to the United States.[56] Investment in the interior layer of the architecture has arisen mainly through historical programs designed to protect and safeguard national nuclear facilities and laboratories.

Congress might expand or reduce agency funding levels to more closely match the levels determined by DNDO to meet the needs of the global nuclear detection architecture, increase overall funding for all aspects of the global nuclear detection architecture to increase redundancy, or decrease funding if it believes other funding priorities are more important. Shifting funding between layers of the architecture has complex ramifications: it may imperil international agreements, lead to perceptions of weakness or strength in the various programs, or cause interagency disagreements. Additionally, unless a robust evaluative system has been established for the global nuclear detection architecture with clear metrics, tying architecture performance to program funding, changes in investment in the different layers of the global nuclear detection architecture may not yield optimal risk reduction. It is difficult to assess without careful evaluation whether shifting funds from one program to another will have a positive or negative net impact; the relative size of the two programs is not necessarily the relevant criterion for assessing its impact on the global nuclear detection architecture. Moreover, DNDO is not statutorily empowered to direct changes in the funding of other agencies. Only through higher-level budgetary policy decisions can interagency funding profiles be changed. This situation may result in a mismatch between the optimal investment levels for the global nuclear detection architecture and the actual investments made. Congress might choose to provide the DNDO Director with the authority to review and assess the budgets of other departments and agencies involved in the global nuclear detection architecture and to comment or recommend alternative budget allocation to the Interagency Coordinating Council or directly to the Office of Management and Budget.[57]

Another possible approach would be for Congress to require that agencies create a single global nuclear detection architecture budget, to provide Congress with a more transparent correlation between agency funding and the global nuclear detection architecture. For example, an annual budget supplement is issued for the National Nanotechnology Initiative, another multi-agency federal endeavor with a large budget. Such a budget supplement for nuclear detection might be coordinated by DNDO through the Interagency Coordinating

Council, Homeland Security Council, National Security Council, or National Science and Technology Council, or through one of the agencies participating in the global nuclear detection architecture. Alternately, Congress might vest a budgetary coordinating role within the Office of Science and Technology Policy rather than within DNDO.

Congress established an Office of the United States Coordinator for the Prevention of Weapons of Mass Destruction Proliferation and Terrorism within the White House. The head of this office is to advise the President, formulate national strategy and policy, lead interagency coordination and implementation, and oversee development of a comprehensive, coordinated budget for weapon of mass destruction proliferation and terrorism prevention.[58] While the scope of the Coordinator's responsibilities is broader than nuclear and radiological issues, the comprehensive, coordinated budget so developed might be one mechanism for clarification of priorities for nuclear detection investment. As of the writing of this report, the position of Coordinator has not been filled.

The affected agencies may not be supportive of the establishment of a unified budget or additional review of agency budget decisions. Reportedly, agencies resisted similar options that were considered during the creation of DNDO, leading to "major limits on both its scope and its power."[59] Agencies may perceive a bottom-up process to be more effective in meeting agency and national needs than a top-down process.

Congress has mandated an annual interagency review of the global nuclear detection architecture. This review is to be overseen by the Secretaries of Homeland Security, State, Defense, and Energy, the Attorney General, and the Director of National Intelligence. Its purpose is to ensure that each participating agency assesses and evaluates its participation in the global nuclear detection architecture. The review is to include evaluation of detection technologies, identification of deficiencies, and assessment of agency capacity for implementation of its responsibilities within the global nuclear detection architecture.[60] This interagency review process maycause the agencies involved to clarify their priorities and funding requirements and thereby cause further evolution of the global nuclear detection architecture.

Balance between Incremental and Transformational Changes to the Global Nuclear Detection Architecture

The DNDO aims to improve "the probability of detection by integrating and deploying current technologies, continually improving these technologies through both near-term enhancements and transformational research and development, and expanding detection capabilities at the Federal, State and local levels."[61] In expanding and improving the global nuclear detection architecture, DNDO and other participating agencies are faced with a temporal choice. Vulnerabilities and gaps identified through the global nuclear detection architecture could be reduced by applying immediately available technologies that provide a partial solution or by investing in research and development to develop technologies that will provide a more complete solution in the long-term. In the first case, the abilities of the global nuclear detection architecture would be incrementally improved as technologies that marginally increase the detection capabilities of the existing architecture are adopted and then serially replaced. Such a strategy might be costly, as multiple generations of equipment, each with some advantage over the previous version, are purchased and deployed. Although each generation would be an improvement, it would not provide a fully acceptable level of detection and security. In the other case, known vulnerabilities might not immediately be addressed at all, allowing the possibility that attackers could exploit them while a research and development program attempted to develop a single system that would remove the vulnerability. Thus, an appreciable risk would remain, even though it could be partly reduced in the near term, until the results of the research and development program came to fruition.

In practice, expansion and improvement of the global nuclear detection architecture requires a balance of these two approaches, using incremental advances and transformational research in coordination to develop a robust architecture. A key concern is how this balance is achieved and identified. The DNDO is addressing this complex problem by developing time-phased plans to address the most important gaps in the existing global nuclear detection architecture.[62] These plans will allow "for the integration of current and near-term technologies and approaches, as well as longer-term options that may draw upon technologies that are currently in the R&D phase and that may not be available for implementation for several years."[63]

Additionally, DNDO has taken a spiral development approach that appears to lean toward deploying available technologies, even if they serve only as partial solutions, rather than leaving a gap unaddressed. In spiral development, technologies are refined as they are implemented based on information generated during their deployment. Thus, though the technology may serve only as a partial solution when first implemented, the goal is for it to become a more complete solution over time.

Finally, DNDO has adopted a multi-faceted approach to architecture development, funding multiple types of detector technology.[64] Successes in developing and deploying these detector types may lead to significant advances in DNDO's ability to detect and prevent a radiological or nuclear attack. In addition to developing and procuring such technologies, DNDO also must weigh the relative advantages to integrating these new systems into the existing architecture versus replacing existing systems with these new systems.

Among key issues facing Congress are determining the optimal process for creating a robust global nuclear detection architecture,understanding the capabilities of near- and long-term technology and their potential effect on the global nuclear detection architecture, and assessing the adequacy of the metrics used tomeasure the risk reduction benefits. Congress bestowed upon DNDO the responsibility for developing and implementing a global nuclear detection architecture as part of efforts to safeguard the United States. The success of DNDO's activities in establishing this architecture will likely require ongoing evaluation and oversight into the future.

Long-Term Maintenance of the Global Nuclear Detection Architecture

Although the use of detailees, IPAs, and liaisons from other agencies has helped DNDO to maintain contact with other stakeholders, its reliance on these temporary personnel may make long-term efforts difficult to sustain. As temporary personnel return to their home agency, institutional memory may be lost. As of April 2007, 79% of DNDO's employees were either detailees, liaisons, or contractors rather than permanent staff.[65] Efforts that rely on continued improvement and adjustment over time, as the global nuclear detection architecture does, will likely depend on DNDO's ability to clearly enunciate and document the rationale and approaches that it developed and considered when they were established. Otherwise, these efforts may be delayed as new personnel have to reevaluate the ongoing process. Similarly, coordination between agencies through the Interagency Coordinating Council may suffer if consensus decisions are not well understood by the successors of the Council participants.

The DNDO may be able to offset this potential loss of institutional memory in a number of ways. One possibility is a mentoring process in which outgoing personnel actively mentor their replacements during an overlap period in order to provide continuity of information and expertise. Another would involve comprehensive documentation of decisions, both positive and negative, so that future staff have a written record to refer to when trying to understand why a particular approach was taken and why a competing approach was set aside. Finally, expanding DNDO's permanent staff might provide long-term stability and more retention of core knowledge.

The DNDO does appear to be increasing its permanent federal staff. As seen in **Table 1**, the number of detailees has decreased and the number of permanent staff has increased since the creation of the office.

Some positions within DNDO may be best filled by permanent DNDO staff, while others may require the expertise possessed only by detailees. A key issue facing decision-makers is balancing DNDO's need for technical or subject matter experts with building a core of permanent DNDO staff able to develop and evolve the nascent global nuclear detection architecture.

The DNDO is aware of the need to retain institutional knowledge. Succession planning is one mechanism DNDO uses to retain this knowledge. According to DNDO, before each detailee returns to their home agency, their replacement is identified and prepared to take over the departing detailee's responsibilities.[66] The effectiveness of this succession planning and DNDO's strategic decisions regarding the use of detailees mayrequire oversight from Congress to ensure that congressional interests in program and agency continuity are met.

Table 1. DNDO Staff Levels

	FY2006 (estimate)	**FY2007 (estimate)**	**FY2008 (estimate)**	**FY2009 (requested)**
Detailees	66	65	45	45-50[a]
Total Staff	112	130	130	144

Source: CRS analysis of DNDO congressional justifications for FY2007, FY2008, and FY2009.
[a] DNDO reported a range in the number of detailees projected for FY2009.

Research and Development Coordination

Research and development investment plays a role in strategies for addressing the architecture's gaps and vulnerabilities. The aim is to develop technologies to fill the gaps in the global nuclear detection architecture. Both DHS and DOE fund research and development in the area of nuclear and radiological detection equipment.

The SAFE Port Act of 2002 gave DNDO the statutory authority to

> conduct, support, coordinate, and encourage an aggressive, expedited, evolutionary, and transformational program of research and development to generate and improve technologies to detect and prevent the illicit entry, transport, assembly, or potential use within the United States of a nuclear explosive device or fissile or radiological material, and coordinate with the Under Secretary for Science and Technology on basic and advanced or transformational research and development efforts relevant to the mission of both organizations.[67]

The research and development activities DNDO undertakes under this authority addresses gaps and vulnerabilities in the global nuclear detection architecture. The DNDO has highlighted detecting threat materials from greater distances, in highly cluttered backgrounds, and in the presence of shielding and masking materials as particular challenges.[68]

Although Homeland Security Presidential Directive 14 required DNDO to "conduct, support, coordinate, and encourage an aggressive, expedited, evolutionary, and transformational program of research and development efforts," it also directed the Secretary of Energy to "lead the development of nonproliferation research and development and, where appropriate, make available dual-use counter-proliferation and counter-terrorism nuclear detection research and development to DNDO and other entities and officials to support the development of the domestic nuclear and radiological detection system."[69] The long-term coordination of this research appears to be occurring through the Subcommittee on Nuclear Defense Research and Development of the National Science and Technology Council Committee on Homeland Defense and National Security.[70] Additional coordination occurs between agencies participating in the global nuclear detection architecture as well.

The coordination of these research and development activities is likely to remain of interestinCongress, toprevent duplication of effort and ensure that agencies meet their missions and roles. Congress may use its oversight authority to assess the balance of investments between agencies, address undue duplication of research and development activities, and increase or decrease the resources available for particular technology approaches under consideration.

End Notes

[1] The term "special nuclear material" was defined by the Atomic Energy Act and includes plutonium and uranium enriched in the isotope 233 or in the isotope 235. See 42 U.S.C. 2014.

[2] For more information about the International Nuclear Materials Protection and Cooperation program, see online at [http://nnsa.energy.gov/nuclear_nonproliferation/ Office%20of%20Int'l%20Material%20Protection%20&%20 Cooperation.htm]. See also CRS Report RL31957, *Nonproliferation and Threat Reduction Assistance: U.S. Programs in the Former Soviet Union*, by Amy F. Woolf.

[3] For more information on DTRA activities in Cooperative Threat Reduction, see online at [http://www.dtra.mil/oe/ctr/programs/index.cfm]. See also CRS Report RL31957, *Nonproliferation and Threat Reduction Assistance*, by Amy F. Woolf

[4] For more information on the Second Line of Defense program, see Office of the Second Line of Defense, National Nuclear Security Administration, *SLD Implementation Strategy: Revision B*, April 2006, online at [http://www.doeal.gov/dicce/ImplementationDocs/ SLDImplentationStrategy.pdf], and the Office of the Second Line of Defense, National Nuclear Security Administration, *Strategic Plan*, 2006, online at [http://www.doeal.gov/ dicce/ImplementationDocs/StrategicPlan.pdf]. See also CRS Report RL31957, *Nonproliferation and Threat Reduction Assistance*, by Amy F. Woolf.

[5] For more information on the Export Control and Related Border Security Assistance Program, see online at [http://www.state.gov/t/isn/export/ecc/20779.htm]. See also CRS Report RL31957, *Nonproliferation and Threat Reduction Assistance*, by Amy F. Woolf.

[6] See, for example, Government Accountability Office, *Combating Nuclear Smuggling: Efforts to Deploy Radiation Detection Equipment in the United States and in Other Countries*, GAO-05-840T, June 21, 2005.

[7] See Defense Science Board, *Report of the Defense Science Board Task Force on Preventing and Defending Against Clandestine Nuclear Attack*, June 2004, and Defense Science Board, *Protecting the Homeland: Report of the Defense Science Board 2000 Summer Study: Executive Summary, Volume I*, February 2001.

[8] Executive Office of the President, The White House, *Domestic Nuclear Detection*, Homeland Security Presidential Directive HSPD-14/National Security Presidential Directive NSPD-43, April 15, 2005.

[9] 6 U.S.C. 592(a)(4).

[10] Domestic Nuclear Detection Office, Department of Homeland Security, *Congressional Justification FY2009*, p. DNDO RD&O-2.

[11] This estimate may understate the total investment, as programs for which nuclear detection is only a component were prorated. This prorating may have not accurately captured the true investment for the nuclear detection component of the program. (Office of Inspector General, Department of Homeland Security, *DHS' Domestic Nuclear Detection Office: Progress in Integrating Detection Capabilities and Response Protocols*, OIG-08-19, December 2007.)

[12] Such as the Container Security Initiative, the Secure Freight Initiative, and the Radiation Portal Monitor program.

[13] Such as the DOD Cooperative Threat Reduction activities.

[14] Such as the DOE Second Line of Defense programs.

[15] Domestic Nuclear Detection Office, Department of Homeland Security, *Congressional Justification FY2008*, p. DNDO RD&O-5.

[16] Government Accountability Office, *Combating Nuclear Smuggling: Efforts to Deploy Radiation Detection Equipment in the United States and in Other Countries*, GAO-05-840T, June 21, 2005.

[17] James Goodby, Timothy Coffey, and Cheryl Loeb, Center for Technology and National Security Policy, National Defense University, *Deploying Nuclear Detection Systems: A Proposed Strategy for Combating Nuclear Terrorism*, July 2007.

[18] Defense Science Board, *Report of the Defense Science Board Task Force on Preventing and Defending Against Clandestine Nuclear Attack*, June 2004, and *Protecting the Homeland: Report of the Defense Science Board 2000 Summer Study: Executive Summary, Volume I*, February 2001.

[19] Domestic Nuclear Detection Office, Department of Homeland Security, *Congressional Justification FY2009*, p. DNDO ACQ-8.

[20] Domestic Nuclear Detection Office, Department of Homeland Security, *Congressional Justification FY2008*, p. DNDO ACQ-7.

[21] See Government Accountability Office, *Combating Nuclear Smuggling: Efforts to Deploy Radiation Detection Equipment in the United States and in Other Countries*, GAO-05-840T, June 21, 2005.

[22] Office of Inspector General, Department of Homeland Security, *DHS' Domestic Nuclear Detection Office: Progress in Integrating Detection Capabilities and Response Protocols*, OIG-08-19, December 2007.

[23] The DNDO states that the largest gaps in their border layer are air, maritime, and land pathways between designated ports of entry. (Domestic Nuclear Detection Office, Department of Homeland Security, *Congressional Justification FY2009*, p. DNDO RD&O-8.)

[24] Domestic Nuclear Detection Office, Department of Homeland Security, *Congressional Justification FY2008*, p. DNDO RD&O-7. The FY2009 congressional justification does not state that DNDO performed an independent peer review of the employed methodology.

[25] Homeland Security Advanced Research Projects Agency, S&T Directorate, Department of Homeland Security, "Radiological and Nuclear Countermeasure System Architectures Analysis (RNCSAA)," BAA 04-01, February 2004.

[26] For an overview of the goals of this study, see Department of Homeland Security, *National Capital Region Coordination: First Annual Report*, September 2005, p. F-41, online at [http://www.dhs.gov/xlibrary/assets/NCR_DHS_Congressional_Final_ 090105.pdf].

[27] Office of Inspector General, *Department of Homeland Security, DHS' Domestic Nuclear Detection Office: Progress in Integrating Detection Capabilities and Response Protocols*, OIG-08-19, December 2007.

[28] James Goodby, Timothy Coffey, and Cheryl Loeb, Center for Technology and National Security Policy, National Defense University, *Deploying Nuclear Detection Systems: A Proposed Strategy for Combating Nuclear Terrorism*, July 2007.

[29] An example of misdirection might be the deployment of decoy radiation detectors. While decoys would have no radiation detection capability, opponents would not know this. They might choose a more risky approach because of their inaccurate understanding of actual detector deployment.

[30] A false positive occurs when the system indicates a threat even though no threat is present.

[31] For the purposes of this report, vulnerabilities refer to when current capabilities are at least partly insufficient to meet current needs. Gaps refer to when no current capability exists to meet a current need.

[32] The DNDO refers to both the absence of detection capability and limitations in existing detection systems as gaps. See Domestic Nuclear Detection Office, Department ofHomeland Security, *Congressional Justification FY2009*, p. DNDO R&DO-8.

[33] The DNDO acknowledges that "no single layer of protection can ever be one hundred percent successful." (Domestic Nuclear Detection Office, Department of Homeland Security, *Congressional Justification FY2009*, p. DNDO ACQ-8.)

[34] The DNDO undertakes a series ofnetassessments, including "red teaming" to identify the effectiveness of the planned and deployed global nuclear detection architecture. (Domestic Nuclear Detection Office, Department of Homeland Security, *Congressional Justification FY2008*, p. DNDO R&DO-2.)

[35] For example, individuals have been able to successfully smuggle surrogate materials into the United States past radiation detection equipment deployed by DHS. Thomas B. Cochran and Matthew G. McKinzie, "Detecting Nuclear Smuggling," *Scientific American*, March 2008.

[36] Domestic Nuclear Detection Office, Department of Homeland Security, *Congressional Justification FY2008*, p. DNDO R&DO-24.

[37] Domestic Nuclear Detection Office, Department of Homeland Security, *Congressional Justification FY2008*, p. DNDO R&DO-8.

[38] See, for example, Government Accountability Office, *Combating Nuclear Smuggling: DHS's Cost-Benefit Analysis to Support the Purchase of New Radiation Detection Portal Monitors Was Not Based on Available Performance Data and Did Not Fully Evaluate All the Monitors' Costs and Benefits*, GAO-07-133R, October 17, 2006.

[39] The DNDO states that its risk-based cost-benefit analysis methodology is used to quantify and prioritize architecture improvements. Domestic Nuclear Detection Office, Department of Homeland Security, *Congressional Justification FY2008*, p. DNDO R&DO-9.

[40] According to the DNDO Inspector General, DNDO has been able to develop a list of detection priorities. (Office of Inspector General, Department of Homeland Security, *DHS' Domestic Nuclear Detection Office: Progress in Integrating Detection Capabilities and Response Protocols*, OIG-08-19, December 2007.) It is unclear how these priorities have been ordered.

[41] Domestic Nuclear Detection Office, Department of Homeland Security, *Congressional Justification FY2008*, p. DNDO RD&O-9.

[42] For example, the methodology underpinning the DHS Bioterrorism Risk Assessment underwent review through the National Academies of Science even though the results of this risk assessment are classified. See National Research Council, *Interim Report on Methodological Improvements to the Department of Homeland Security's Biological Agent Risk Analysis*, National Academies Press, 2007.

[43] Office of Inspector General, Department of Homeland Security, *DHS' Domestic Nuclear Detection Office: Progress in Integrating Detection Capabilities and Response Protocols*, OIG-08-19, December 2007.

[44] 6 U.S.C. 591(a).

[45] The Intergovernmental Personnel Act allows for the temporary transfer of personnel to a federal agency. See 5 U.S.C. 3371-76.

[46] Domestic Nuclear Detection Office, Department of Homeland Security, *Congressional Justification FY2009*, p. DNDO-1.

[47] Office of Inspector General, Department of Homeland Security, *DHS' Domestic Nuclear Detection Office: Progress in Integrating Detection Capabilities and Response Protocols*, OIG-08-19, December 2007.

[48] A similar situation exists with the requirement for the DHS Under Secretary for Science and Technology to coordinate homeland security research and development across the federal government. In this case, the Under Secretary for Science and Technology was able to release a coordination document without the consensus of other government agencies, but rather with other agency consultation.

[49] Office of Inspector General, Department of Homeland Security, *DHS' Domestic Nuclear Detection Office: Progress in Integrating Detection Capabilities and Response Protocols*, OIG-08-19, December 2007.

[50] Office of Inspector General, Department of Homeland Security, *DHS' Domestic Nuclear Detection Office: Progress in Integrating Detection Capabilities and Response Protocols*, OIG-08-19, December 2007.

[51] Testimony of DNDO Director Vayl S. Oxford, Department of Homeland Security, before the House Homeland Security Committee, Subcommittee on Emerging Threats, Cybersecurity, and Science and Technology, on October 10, 2007.

[52] P.L. 109-295, Title IV.

[53] Department of Homeland Security, *Congressional Justification FY2009*, p. DNDO M&A-4.

[54] For more information on the Cooperative Threat Reduction program, see CRS Report RL31957, *Nonproliferation and Threat Reduction Assistance: U.S. Programs in the FormerSoviet Union*, by Amy F. Woolf.

[55] The Advanced Spectroscopic Portal is a radiation detection system designed to be able to identify the source of the detected radiation. The Cargo Advanced Automated Radiography System is an imaging system designed to automatically detect special nuclear material by X-ray imaging. The Advanced Spectroscopic Portal technology has been the subject of a number of congressional hearings and Government Accountability Office audits. See, for example, House Committee on Homeland Security, Subcommittee on Emerging Threats,

Cybersecurity, and Science and Technology, "Nuclear Smuggling Detection: Recent Tests of Advanced Spectroscopic Portal Monitors," hearing held March 5, 2008; and House Committee on Energy and Commerce, Subcommittee on Oversight and Investigations, "Nuclear Terrorism Prevention: Status Report on the Federal Government's Assessment of New Radiation Detection Monitors," hearing held September 18, 2007. See also, Government Accountability Office, *Combating Nuclear Smuggling: Additional Actions Needed to Ensure Adequate Testing of Next Generation Radiation Detection Equipment*, GAO-07-1247T, September 18, 2007; and Government Accountability Office, *Combating Nuclear Smuggling: DHS's Decision to Procure and Deploy the Next Generation of Radiation Detection Equipment Is Not Supported by Its Cost-Benefit Analysis*, GAO-07-581T, March 14, 2007.

[56] Section 1701 of the ImplementingRecommendations of the 9/11 Commission Act of 2007 (P.L. 110-53) requires that, by July 1, 2012, all maritime cargo containers entering the United States be scanned by nonintrusive imaging equipment and radiation detection equipment at a foreign port before departure.

[57] The Director of the National Security Agency has a similar type of authority for national security telecommunications and information systems security programs. Executive Office of the President, The White House, *National Policy for the Security of National Security Telecommunications and Information Systems*, National Security Directive 42 (NSD-42), July 5, 1990.

[58] P.L. 110-53, Title XVII, Subtitle D, Sec. 1841.

[59] Michael A. Levi, *On Nuclear Terrorism*, Harvard University Press, November 2007, p. 146.

[60] P.L. 110-53, Title XI, Sec. 1103.

[61] Domestic Nuclear Detection Office, Department of Homeland Security, *Congressional Justification FY2008*, p. DNDO RD&O-8.

[62] Domestic Nuclear Detection Office, Department of Homeland Security, *Congressional Justification FY2009*, p. DNDO RD&O-7.

[63] Domestic Nuclear Detection Office, Department of Homeland Security, *Congressional Justification FY2009*, p. DNDO RD&O-8.

[64] For example, the DNDO has invested in both portal and portable radiation detector technology as well as radiography system development.

[65] Office of Inspector General, Department of Homeland Security, *DHS' Domestic Nuclear Detection Office: Progress in Integrating Detection Capabilities and Response Protocols*, OIG-08-19, December 2007.

[66] Oral testimony of Mark Mullen, Assistant Director for Architecture, Domestic Nuclear Detection Office, before the Senate Homeland Security and Governmental Affairs Committee on July 16, 2008.

[67] 6 U.S.C. 592.

[68] Testimony of DNDO Director Vayl S. Oxford, Department of Homeland Security, before the Senate Judiciary Committee, Subcommittee on Terrorism, Technology, and Homeland Security, on July 27, 2006.

[69] Executive Office of the President, The White House, *Domestic Nuclear Detection*, Homeland Security Presidential Directive HSPD-14/National Security Presidential Directive NSPD-43, April 15, 2005.

[70] Testimony of DNDO Director Vayl S. Oxford, Department of Homeland Security, before the House Homeland Security Committee, Subcommittee on Emerging Threats, Cybersecurity, and Science and Technology, on October 10, 2007.

In: Detection of Nuclear Weapons and Materials
Editor: Pietro Egidi
ISBN: 978-1-60741-512-1

Chapter 4

NUCLEAR DETECTION: PRELIMINARY OBSERVATIONS ON THE DOMESTIC NUCLEAR DETECTION OFFICE'S EFFORTS TO DEVELOP A GLOBAL NUCLEAR DETECTION ARCHITECTURE

Government Accountability Office

WHY GAO DID THIS STUDY

In April 2005, a Presidential Directive established the Domestic Nuclear Detection Office (DNDO) within the Department of Homeland Security to enhance and coordinate federal, state, and local efforts to combat nuclear smuggling domestically and abroad. DNDO was directed to develop, in coordination with the departments of Defense (DOD), Energy (DOE), and State (State), an enhanced global nuclear detection architecture—an integrated system of radiation detection equipment and interdiction activities. DNDO implements the domestic portion of the architecture, while DOD, DOE, and State are responsible for related programs outside the U.S.

This testimony provides preliminary observations based on ongoing work addressing (1) the status of DNDO's efforts to develop a global nuclear detection architecture, (2) the challenges DNDO and other federal agencies face in implementing the architecture, and (3) the costs of the programs that constitute the architecture. This statement draws on prior GAO reviews of programs constituting the architecture, and GAO's work on strategic planning.

WHAT GAO RECOMMENDS

GAO recommends that DNDO develop, in coordination with DOD, DOE, and State, a strategic plan to guide agency efforts to develop a more comprehensive architecture. In commenting on a draft of this statement, DNDO concurred with this recommendation.

To view the full product, including the scope and methodology, click on GAO-08-999T. For more information, contact David Maurer at (202) 512-3841 or maurerd@gao.gov.

WHAT GAO FOUND

According to GAO's preliminary work to date, DNDO has taken steps to develop a global nuclear detection architecture but lacks an overarching strategic plan to help guide how it will achieve a more comprehensive architecture. Specifically, DNDO has developed an initial architecture after coordinating with DOD, DOE, and State to identify 74 federal programs that combat smuggling of nuclear or radiological material. DNDO has also identified gaps in the architecture, such as land border crossings into the United States between formal points of entry, small maritime vessels, and international general aviation. Although DNDO has started to develop programs to address these gaps, it has not yet developed an overarching strategic plan to guide its transition from the initial architecture to a more comprehensive architecture. For example, such a plan would define across the entire architecture how DNDO would achieve and monitor its goal of detecting the movement of radiological and nuclear materials through potential smuggling routes, such as small maritime craft or land borders in between points of entry. The plan would also define the steps and resources needed to achieve a more comprehensive architecture and provide metrics for measuring progress toward goals.

DNDO and other federal agencies face a number of coordination, technological, and management challenges. First, prior GAO reports have demonstrated that U.S.-funded radiological detection programs overseas have proven problematic to implement and sustain and have not been effectively coordinated, although there have been some improvements in this area. Second, detection technology has limitations and cannot detect and identify all radiological and nuclear materials. For example, smugglers may be able to effectively mask or shield radiological materials so that it evades detection. Third, DNDO faces challenges in managing implementation of the architecture. DNDO has been charged with developing an architecture that depends on programs implemented by other agencies. This responsibility poses a challenge for DNDO in ensuring that the individual programs within the global architecture are effectively integrated and coordinated to maximize the detection and interdiction of radiological or nuclear material.

According to DNDO, approximately $2.8 billion was budgeted in fiscal year 2007 for the 74 programs included in the global nuclear detection architecture. Of this $2.8 billion, $1.1 billion was budgeted for programs to combat nuclear smuggling internationally; $220 million was devoted to programs to support the detection of radiological and nuclear material at the U.S. border; $900 million funded security and detection activities within the United States; and approximately $575 million was used to fund a number of cross-cutting activities. The future costs for DNDO and other federal agencies to address the gaps identified in the initial architecture are not yet known or included in these amounts.

Mr. Chairman and Members of the Committee:

I am pleased to be here today to discuss our preliminary work for several members of Congress on the Domestic Nuclear Detection Office's (DNDO) efforts to develop a global nuclear detection architecture—essentially, an integrated system of radiation detection equipment and interdiction activities to combat nuclear smuggling in foreign countries, at the U.S. border, and inside the United States. Preventing terrorists from using radiological or nuclear material to carry out an attack in the United States is a top national priority. Since the events of September 11, 2001, there is heightened concern that terrorists may try to smuggle nuclear materials or a nuclear weapon into the United States. If terrorists were to carry out such an attack, the consequences could be devastating to our national interests.

In April 2005, the President issued a directive establishing DNDO, within the Department of Homeland Security (DHS), to enhance and coordinate federal, state, and local efforts to prevent radiological and nuclear attacks.[1] Congress subsequently passed the SAFE Port Act of 2006, which established DNDO in statute.[2] Among other things, DNDO must develop, in coordination with the departments of Defense (DOD), Energy (DOE), and State (State), an enhanced global nuclear detection architecture. DNDO is explicitly charged with implementing the domestic portion (at the U.S. border and within the United States) of the architecture and with coordinating the nuclear detection efforts of federal, state, and local governments. It is also responsible for developing, acquiring, and deploying radiation detection equipment to support the efforts of DHS and other federal agencies. The directive and the SAFE Port Act also reaffirmed that DOD, DOE, and State are responsible for programs to combat radiological and nuclear smuggling outside the United States.

Over the past few months, we have been examining the steps that DNDO has taken to develop a global nuclear detection architecture. Our work is ongoing and our statement today will provide preliminary observations on DNDO's effort. Specifically, our statement will discuss (1) the status of DNDO's efforts to develop a global nuclear detection architecture, (2) the challenges DNDO and the other federal agencies face in implementing the architecture, and (3) the costs of the current and proposed programs that constitute the global nuclear detection architecture. We plan to issue our final report in January 2009.

To begin addressing these objectives, we interviewed officials from DNDO about steps taken to develop and improve upon the existing architecture. We reviewed and analyzed documents DNDO used to help create the baseline, or initial, architecture, as well as studies on gaps identified in the architecture.[3] We interviewed agency officials from DOD, DOE, and State who manage programs that are part of the architecture to get their perspectives on challenges faced in implementing the architecture. In addition, we interviewed subject matter experts from the academic and nonprofit sectors, as well as representatives from the International Atomic Energy Agency (IAEA), to gain their perspective on efforts to develop and implement the architecture. We have conducted our work to date in accordance with generally accepted government auditing standards. Those standards require that we plan and perform the audit to obtain sufficient, appropriate evidence to provide a reasonable basis for our findings and conclusions based on our audit objectives. We believe that the evidence obtained provides a reasonable basis for our findings and conclusions based on our audit objectives.

SUMMARY

According to our preliminary work to date, DNDO has taken steps to develop a global nuclear detection architecture, but it lacks an overarching strategic plan to help guide how it will achieve a more comprehensive architecture. Specifically, DNDO has developed an initial architecture after coordinating with, among others, DOD, DOE, and State, to identify 74 federal programs that combat smuggling of nuclear or radiological material. Many of these programs predate the establishment of DNDO. These programs cover all of the layers of detection, including securing special nuclear and radiological materials at their source in foreign countries and in the United States as well as detecting these materials at U.S. borders or within the United States. DNDO has also collaborated with other federal agencies, such as DOD, DOE, and State, to (1) identify gaps in the initial architecture, such as land borders between ports of entry, small maritime vessels, and international general aviation, and (2) develop programs to address these gaps. For example, DNDO has a joint project underway with the Customs and Border Protection's (CBP) Office of Border Patrol to assess the feasibility of equipping border patrol agents with portable radiological and nuclear detection equipment along the U.S. border. Although these efforts to address recognized gaps in the architecture are necessary first steps, DNDO has not developed an overarching strategic plan that will guide its transition from the initial architecture to a more comprehensive architecture. Such a plan would define across the entire architecture how, for example, DNDO will achieve its goal of detecting the movement of radiological and nuclear materials through potential smuggling routes, such as small maritime craft or land borders in between ports of entry. This plan would also define the steps and resources needed to achieve a more comprehensive architecture and provide metrics for measuring progress toward goals, such as enhancing detection along borders.

DNDO and other federal agencies face a number of coordination, technological, and management challenges in developing a more comprehensive detection architecture. First, prior GAO reports have demonstrated that critical, long-standing U.S.-funded radiological detection programs overseas—which are a critical component of the architecture—have proven problematic to implement and sustain and have not been effectively coordinated. Although coordination among the agencies has improved, as a chain is only as strong as its weakest link, challenges in any of the programs that constitute the architecture may ultimately limit its overall effectiveness. Second, detection technology has limitations and is currently unable to detect and identify all smuggled radiological and nuclear materials. For example, smugglers may be able to effectively mask or shield radiological materials so that it evades detection. We have also previously raised concerns about DNDO's efforts to develop a new generation of radiation detection equipment. Furthermore, while radiation detection equipment is an important part of the architecture, combating nuclear smuggling requires an integrated approach that also includes proper training and intelligence gathering on smuggling operations. Third, DNDO faces challenges in managing the implementation of the architecture. DNDO has been charged with developing an architecture that is dependent on programs implemented by other agencies. Ensuring that these individual programs within the architecture are effectively integrated poses a challenge for DNDO. In addition, the majority of the employees in DNDO's architecture office are on detail from other federal agencies or are contractors. While this staffing approach taps expertise across several agencies, detailees return to their home organizations after a relatively short time and in some cases there have been delays in filling these vacancies. As a result, this turnover may limit the retention and depth of institutional memory.

According to DNDO, approximately $2.8 billion was budgeted in fiscal year 2007 for the 74 programs included in the global nuclear detection architecture. Of this $2.8 billion, approximately $1.1 billion was budgeted for programs designed to combat nuclear smuggling and secure materials internationally. Approximately $220 million was devoted to programs to support the detection of radiological and nuclear material at the U.S. border; and an additional $900 million funded security and detection activities within the United States. Finally, approximately $575 million was used to fund a number of cross-cutting activities that support many different layers of the architecture, such as those focused on research and development or technical support to users of the detection equipment. In addition to these programs, DNDO and other federal agencies are pursuing future initiatives to address the gaps identified in the initial architecture. The costs to implement and sustain these and other future programs and equipment are not yet known or included in these figures. We are in the process of reviewing this cost information and will provide more detailed analysis in our final report.

BACKGROUND

According to IAEA, between 1993 and 2006, there were 1,080 confirmed incidents of illicit trafficking and unauthorized activities involving nuclear and radiological materials worldwide. Eighteen of these cases involved weapons-usable material—plutonium and highly enriched uranium (HEU)—that could be used to produce a nuclear weapon. IAEA also reported that 124 cases involved materials that could be used to produce a device that uses conventional explosives with radioactive material (known as a "dirty bomb"). Past confirmed incidents of illicit trafficking in HEU and plutonium involved seizures of kilogram quantities of weapons-usable nuclear material but most have involved very small quantities. In some of these cases, it is possible that the seized material was a sample of larger quantities available for illegal purchase. IAEA concluded that these materials pose a continuous potential security threat to the international community, including the United States.

Nuclear material could be smuggled into the United States in a variety of ways: hidden in a car, train or ship; sent through the mail; carried in personal luggage through an airport; or walked across an unprotected border. In response to these threats, U.S. agencies, including DHS, DOD, DOE, and State, implemented programs to combat nuclear smuggling in foreign countries and the United States. DOD, DOE, and State fund, manage, and implement the global nuclear detection architecture's international programs. Many international detection programs were operating for several years before DNDO was created. For example, DOE's Materials Protection, Control, and Accounting program, initiated in 1995, provides support to the Russian Federation and other countries of concern to secure nuclear weapons and weapons material that may be at risk of theft of diversion. In addition, during the 1990s, the United States began deploying radiation detection equipment at borders in countries of the former Soviet Union. DOD's Cooperative Threat Reduction (CTR) program launched a variety of programs in the early 1990s to help address proliferation concerns in the former Soviet Union, including helping secure Russian nuclear weapons. Two other DOD programs have provided radiation portal monitors, handheld equipment, and radiation detection training to countries in the former Soviet Union and in Eastern Europe. Similarly, State programs have provided detection equipment and training to numerous countries. DHS, in conjunction with other federal, state, and local agencies, is responsible for combating nuclear smuggling in the United States and has provided radiation detection equipment, including portal monitors, personal radiation detectors (known as pagers), and radioactive isotope identifiers at U.S. ports of entry.

All radiation detection devices have limitations in their ability to detect and identify nuclear material. Detecting attempted nuclear smuggling is difficult because there are many sources of radiation that are legal and not harmful when used as intended. These materials can trigger alarms—known as nuisance alarms—that may be indistinguishable in some cases from alarms that could sound in the event of a true case of nuclear smuggling. Nuisance alarms can be caused by patients who have recently had cancer treatments; a wide range of cargo with naturally occurring radiation (e.g., fertilizer, ceramics, and food products) and legitimate shipments of radiological sources for use in medicine and industry.

DNDO Has Begun to Develop Programs to Enhance the Initial Architecture, but It Lacks an Overarching Strategic Plan for Achieving Future Architecture Improvements

In October 2005, a few months after its inception, DNDO completed its initial inventory of federal programs associated with detecting the illicit transport of radiological and nuclear materials. As part of this effort, DNDO defined the architecture's general approach: a multilayered detection framework of radiation detection equipment and interdiction activities to combat nuclear smuggling in foreign countries, at the U.S. border, and inside the United States. DNDO, in collaboration with other federal agencies, such as DOD, DOE, and State, analyzed the gaps in current planning and deployment strategies to determine the ability of individual layers of the architecture to successfully prevent illicit movement of radiological or nuclear materials or devices. DNDO identified several gap areas with respect to detecting potential nuclear smuggling, such as (1) land border crossings into the United States between formal points of entry, (2) small maritime craft (any vessel less than 300 gross tons) that enter the United States, and (3) international general aviation.

In November 2006, DNDO completed a more detailed analysis of programs in the initial architecture. DNDO identified 72 programs across the federal government that focused on combating radiological and nuclear smuggling and nuclear security and it discussed these programs in depth by layer. The analysis also included a discussion of the current and anticipated budgets associated with each of these programs and each of the layers. In June 2008, DNDO released the *Joint Annual Interagency Review of the Global Nuclear Detection Architecture*. This report provides an updated analysis of the architecture by layer of defense and a discussion of the 74 programs now associated with each of the layers, as well as an estimate of the total budgets by layer.

To address the gaps identified in the domestic portions of the architecture, DNDO has initiated pilot programs to address primary areas of concern or potential vulnerability. For example:

- For the land border in between ports of entry, DNDO and CBP are studying the feasibility of equipping CBP border patrol agents with portable radiological and nuclear detection equipment along the U.S. border.
- For small marine vessels, DNDO is working with the Coast Guard to develop and expand the coverage of radiological and nuclear detection capabilities that can be specifically applied in a maritime environment.
- For international general aviation, DNDO is working with CBP, the Transportation Security Administration, and other agencies to develop and implement radiological and nuclear detection capabilities to scan international general aviation flights to the United States for possible illicit radiological or nuclear materials.

To date, we have received briefings on each of these programs from DNDO, but we have not yet fully reviewed how they are being implemented. We will examine each of these more closely during the course of our review.

Our preliminary observation is that DNDO's pilot programs appear to be a step in the right direction for improving the current architecture. However, these efforts to address gaps are not being undertaken within the larger context of an overarching strategic plan. While each agency that has a role in the architecture may have its own planning documents, DNDO has not produced an overarching strategic plan that can guide its efforts to address the gaps and move to a more comprehensive global nuclear detection architecture. Our past work has discussed the importance of strategic planning.[4] Specifically, we have reported that strategic plans should clearly define objectives to be accomplished, identify the roles and responsibilities for meeting each objective, ensure that the funding necessary to achieve the objectives is available, and employ monitoring mechanisms to determine progress and identify needed improvements. For example, such a plan would define how DNDO will achieve and monitor the goal of detecting the movement of radiological and nuclear materials through potential smuggling routes, such as small maritime craft or land borders in between ports of entry. Moreover, this plan would include agreed-upon processes and procedures to guide the improvement of the architecture and coordinate the activities of the participating agencies.

DNDO and Other Agencies Face Coordination, Technological, and Management Challenges

DNDO and other agencies face a number of challenges in developing a global nuclear detection architecture, including (1) coordinating detection efforts across federal, state, and local agencies and with other nations, (2) dealing with the limitations of detection technology, and (3) managing the implementation of the architecture.

Federal Agencies Have Experienced Difficulties Coordinating Radiation Detection Efforts

Our past work on key aspects of international and domestic programs that are part of the architecture have identified a number of weaknesses. In order for the architecture to be effective, all parts need to be well thought out, managed, and coordinated. As a chain is only as strong as its weakest link, limitations in any of the programs that constitute the architecture may ultimately limit its effectiveness. Specifically, in past work, we have identified the following difficulties that federal agencies have had coordinating and implementing radiation detection efforts.

- We reported that DOD, DOE, and State had not coordinated their approaches to enhance other countries' border crossing.[5] Specifically, radiation portal monitors that State installed in more than 20 countries are less sophisticated than those DOD and DOE installed. As a result, some border crossings where U.S. agencies had installed radiation detection equipment were more vulnerable to nuclear smuggling than others.[6] Since issuing our report, a governmentwide plan encompassing U.S. efforts to combat nuclear smuggling in other countries has been developed; duplicative programs have been consolidated; and coordination among the agencies, although still a concern, has improved.
- In 2005, we reported that there is no governmentwide guidance for border security programs that delineates agencies' roles and responsibilities, establishes regular information sharing, and defines procedures for resolving interagency disputes.[7] In the absence of guidance for coordination, officials in some agencies questioned other agencies' roles and responsibilities.
- More recently, in 2008, we found that levels of collaboration between U.S. and host government officials varied at some seaports participating in DHS's Container Security Initiative (CSI).[8] In addition, we identified hurdles to cooperation between CSI teams and their counterparts in the host government, such as a host country's legal restrictions that CBP officials said prevent CSI teams from observing examinations.

  Furthermore, many international nuclear detection programs rely heavily on the host country to maintain and operate the equipment. We have reported that in some instances this reliance has been problematic. For example:
- About half of the portal monitors provided to one country in the former Soviet Union were never installed or were not operational. In additional, mobile vans equipped with radiation detection equipment furnished by State have limited usefulness because they cannot operate effectively in cold climates or are otherwise not suitable for conditions in some countries.[9]
- Once the equipment is deployed, the United States has limited control over it, as we have previously reported.[10] Specifically, once DOE finishes installing radiation equipment at a port and passes control of the equipment to the host government, the United States no longer controls the equipment's specific settings or its use by foreign customs officials. Settings can be changed, which may decreased the probability that the equipment will detect nuclear material.

Within the U.S. borders, DNDO faces coordination challenges and will need to ensure that the problems with nuclear detection programs overseas are not repeated domestically. Many pilot programs DNDO is developing to address gaps in the architecture will rely heavily on other agencies to implement them. For example, DNDO is working closely with the Coast Guard and other federal agencies to implement DNDO's maritime initiatives to enhance detection of radiological and nuclear materials on small vessels. However, maritime jurisdictional responsibilities and activities are shared among federal, state, regional, and local governments. As a result, DNDO will need to closely coordinate activities related to detecting radiological and nuclear materials with these entities, as well as ensure that users are adequately trained and technical support is available. DNDO officials told us they are closely coordinating with other agencies, and our work to assess this coordination is still underway. We will continue to explore these coordination activities and challenges as we continue our review.

Limitations in Detection Technology Hamper the Architecture's Effectiveness

The ability to detect radiological and nuclear materials is a critical component of the global nuclear detection architecture; however, current technology may not able to detect and identify all smuggled radiological and nuclear materials. In our past work, we found limitations with radiation detection equipment.[11] For example:

- In a report on preventing nuclear smuggling, we found that a cargo container containing a radioactive source was not detected as it passed through radiation detection equipment that DOE had installed at a foreign seaport because the radiation emitted from the container was shielded by a large amount of scrap metal. Additionally, detecting actual cases of illicit trafficking in weapons-usable nuclear material is complicated: one of the materials of greatest concern in terms of proliferation—highly enriched uranium—is among the most difficult materials to detect because of its relatively low level of radioactivity.
- We reported that current portal monitors deployed at U.S. borders can detect the presence of radiation but cannot distinguish between harmless radiological materials, such as ceramic tiles, fertilizer, and bananas, and dangerous nuclear materials, such as plutonium and uranium. DNDO is currently testing a new generation of portal monitors. We have raised continuing concerns about DNDO's efforts to develop and test these advanced portal monitors.[12] We currently have additional work underway examining the current round of testing and expect to report on our findings in September 2008.
- Environmental conditions can affect radiation detection equipment's performance and sustainability, as we also have previously reported. For example, wind disturbances can vibrate the equipment and interfere with its ability to detect radiation. In addition, sea spray may corrode radiation detection equipment and its components that are operated in ports or near water. Its corrosive nature, combined

with other conditions such as coral in the water, can accelerate the degradation of equipment.

It is important to note that radiation detection equipment is only one of the tools that customs inspectors and border guards must use to combat nuclear smuggling. Combating nuclear smuggling requires an integrated approach that includes equipment, proper training, and intelligence gathering on smuggling operations. In the past, most known interdictions of weapons-useable nuclear materials have resulted from police investigations rather than by radiation detection equipment installed at border crossings.

DNDO Faces Challenges in Managing the Global Nuclear Detection Architecture

The task DNDO has been given—developing an architecture to keep radiological and nuclear materials from entering the country—is a complex and large undertaking. DNDO has been charged with developing an architecture that depends on programs implemented by other agencies. This lack of control over these programs poses a challenge for DNDO in ensuring that all individual programs within the global nuclear detection architecture will be effectively integrated. Moreover, implementing and sustaining the architecture requires adequate resources and capabilities to meet needed commitments. However, the majority of the employees in DNDO's architecture office are detailees on rotation from other federal agencies or are contractors. This type of staffing approach allows DNDO to tap into other agencies' expertise in radiological and nuclear detection. However, officials told us that staff turnover may limit the retention and depth of institutional memory since detailees return to their home organizations after a relatively short time. In some cases, there have been delays in filling these vacancies. We will continue to examine this potential resource challenge as we complete our work.

In spite of these challenges, DNDO's efforts to develop a global nuclear detection architecture have yielded some benefits, according to DOD, DOE, and State officials. For example, an official from the State Department told us that DNDO is working through State's Global Initiative to Combat Nuclear Terrorism to develop model guidelines that other nations can use to establish their own nuclear detection architectures and recently sponsored a related workshop. In addition, DOE officials said that DNDO's actions have helped broaden their perspective on the deployment of radiation detection equipment overseas. Previously, the U.S. government had been more focused on placing fixed detectors at particular sites, but as a result of DNDO's efforts to identify gaps in the global detection network, DOE has begun to work with law enforcement officials in other countries to improve detection capabilities for the land in between ports of entry. Finally, DNDO, DOD, DOE, and the Office of the Director of National Intelligence for Science and Technology are now formally collaborating on nuclear detection research and development and they have signed a memorandum of understanding (MOU) to guide these efforts. The MOU will integrate research and development programs by, for example, providing open access to research findings in order to leverage this knowledge and to reduce conflict between different agency programs. In addition, the MOU encourages joint funding of programs and projects and calls on the agencies to coordinate their research and development plans. In our ongoing work, we will examine DNDO's progress in carrying through on these initiatives.

Approximately $2.8 Billion in Fiscal Year 2007 Funded Programs Associated with Detecting the Transport of Radiological and Nuclear Weapons or Materials

DNDO reported that approximately $2.8 billion was budgeted in fiscal year 2007 for 74 programs focused on preventing and detecting the illicit transport of radiological or nuclear materials.[13] These programs were primarily administered by DHS, DOD, DOE, and State and spanned all layers of the global nuclear detection architecture. Specifically:

- $1.1 billion funded 28 programs focused on the international aspects of the architecture;
- $221 million funded 9 programs to support detection of radiological and nuclear material at the U.S. border;
- $918 million funded 16 programs dedicated to detecting and securing radiological or nuclear materials within the U.S. borders; and
- $577 million funded 34 cross-cutting programs that support many different layers of the architecture by, for example, research and development or technical support to users of the detection equipment.

The fiscal year 2007 budget of $2.8 billion will not sustain the architecture over the long term because additional programs and equipment will be implemented to address the gaps. For example, this amount does not include the cost estimates related to acquiring and deploying the next generation of advanced portal monitors that are currently being tested. In addition, DNDO is just beginning new efforts to mitigate gaps in the architecture and budget estimates for these activities are limited. We are in the process of reviewing this cost information and will provide more detailed analysis in our final report.

Conclusions

DNDO has been given an important and complex task—develop a global nuclear detection architecture to combat nuclear smuggling and keep radiological and nuclear weapons or materials from entering the United States. This undertaking involves coordinating a vast array of programs and technological resources that are managed by many different agencies and span the globe. Since its creation 3 years ago, DNDO has conceptually mapped the current architecture and identified how it would like the architecture to evolve in the near term. While DNDO's vision of a more comprehensive architecture is laudable, to achieve this goal, it will need to address a number of key challenges including building close coordination and cooperation among the various agencies involved and developing and deploying more advanced radiation detection technology. Although DNDO has taken some steps to achieve these ends, it has not done so within the larger context of an overarching strategic plan with clearly established goals, responsibilities, priorities, resource needs, and mechanisms for assessing progress along the way. Developing and implementing a global nuclear detection architecture will likely take several years, cost billions of dollars, and rely on the expertise and resources of agencies and programs across the government. Moving forward, DNDO should work closely with its counterparts within DHS, as well as at other departments, to develop a comprehensive strategic plan that helps safeguard the investments made to date, more closely links future goals with the resources necessary to achieve those goals, and enhance the architecture's ability to operate in a more cohesive and integrated fashion.

RECOMMENDATIONS FOR EXECUTIVE ACTION

We recommend that the Secretary of Homeland Security, in coordination with the Secretary of Defense, the Secretary of Energy, and the Secretary of State, develop a strategic plan to guide the development of a more comprehensive global nuclear detection architecture. Such a plan should (1) clearly define objectives to be accomplished, (2) identify the roles and responsibilities for meeting each objective, (3) identify the funding necessary to achieve those objectives, and (4) employ monitoring mechanisms to determine programmatic progress and identify needed improvements.

End Notes

[1] Homeland Security Presidential Directive 14/National Security Presidential Directive 43, *Domestic Nuclear Detection*, April 15, 2005.

[2] 6 U.S.C. §§ 591-596a.

[3] To assess DNDO's efforts to develop the architecture, we are in the process of comparing DNDO's planning practices with those that we have found to be effective in developing enterprise architectures. An enterprise architecture is a tool, or blueprint, for understanding and planning complex systems. GAO has developed an Enterprise Architecture Management Maturity Framework. This framework outlines steps toward achieving a stable and mature process for managing the development, maintenance, and implementation of enterprise architectures. See GAO, *Information Technology: A Framework for Assessing and Improving Enterprise Architecture Management (Version 1.1), GAO-03-584G* (Washington, D.C.: April 2003).

[4] GAO, *Managing for Results: Enhancing Agency Use of Performance Information for Management Decision Making*, GAO-05-927 (Washington, D.C.: Sept. 9, 2005); GAO, *Results-Oriented Government: Practices That Can Help Enhance and Sustain Collaboration among Federal Agencies*, GAO-06-15 (Washington, D.C.: Oct. 21, 2005); GAO, *Combating Terrorism: Observations on National Strategies Related to Terrorism*, GAO-03-519T (Washington, D.C.: Mar. 3, 2003); and GAO, *Executive Guide: Effectively Implementing the Government Performance and Results Act*, GAO/GGD-96-118 (Washington, D.C.: June 1996).

[5] GAO, *Combating Nuclear Smuggling: Efforts to Deploy Radiation Detection Equipment in the United States and in Other Countries*, GAO-05-840T (Washington, D.C.: June 21, 2005).

[6] Portal monitors installed by State do not have the ability to detect neutron radiation, which translates into a decreased ability of those monitors to be able to detect plutonium, one of the nuclear materials of greatest proliferation concern.

[7] GAO, *Weapons of Mass Destruction: Nonproliferation Programs Need Better Integration*, GAO-05-157 (Washington, D.C.: Jan. 28, 2005).

[8] GAO, *Supply Chain Security: Examinations of High-Risk Cargo at Foreign Seaports Have Increased, but Improved Data Collection and Performance Measures Are Needed*, GAO-08-187 (Washington, D.C.: Jan. 25, 2008).

[9] GAO-05-840T.

[10] GAO, *Preventing Nuclear Smuggling: DOE Has Made Limited Progress in Installing Radiation Detection Equipment at Highest Priority Seaports*, GAO-05-375 (Washington, D.C.: Mar. 31, 2005).

[11] GAO-05-375.

[12] GAO, *Combating Nuclear Smuggling: Additional Actions Needed to Ensure Adequate Testing of Next Generation Radiation Detection Equipment*, GAO-07-1247T (Washington, D.C.: Sept. 18, 2007).

[13] The total number of programs reported by DNDO as being related to the architecture is 74. However, the sum of the programs by layer is more than 74 because some programs are relevant to more than one layer of detection.

CHAPTER SOURCES

The following chapters have been previously published:

Chapter 1 – This is an edited, excerpted and augmented edition of a United States Government Accountability Office (GAO), Report to Congressional Committees. Publication GAO-08-979, dated September 2008.

Chapter 2 – This is an edited, excerpted and augmented edition of a United States Congressional Research Service publication, Report Order Code R40154, dated November 6, 2008.

Chapter 3 – This is an edited, excerpted and augmented edition of a United States Congressional Research Service publication, Report Order Code RL34574, dated September 23, 2008.

Chapter 4 – These remarks were delivered as Statement of David C. Maurer, Acting Director, Natural Resources and Environment, before the Committee on Homeland Security and Governmental Affairs, U.S. Senate, dated July 16, 2008.

INDEX

C

D

H

I

O

P

R

S

T

U

V

W

X

Y

Z